AGE 11-14

CHARLES LETTS
Letts
FOUNDED 1796

NATIONAL
CURRICULUM

KEY STAGE THRF

CW00348880

SCIENCE

Bob McDuell

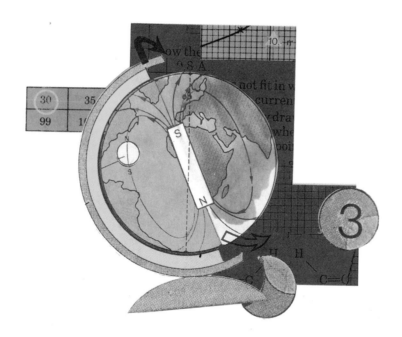

Charles Letts & Co Ltd
London, Edinburgh & New York

First published 1991
by Charles Letts & Co Ltd
Diary House, Borough Road, London SE1 1DW

Editorial team
Angela Royal, Chris Norris

Design team
Anne Davison, Keith Anderson, Peter Holroyd

Text: © Bob McDuell 1991

Illustrations: Michael Renouf, Tek-Art, Artistic License

© Charles Letts & Co Ltd

**British Library Cataloguing in
Publication Data**
McDuell, G. R. (Godfrey Robert) 1944-
Science.
Key stage 3.
1. Science
500

ISBN 0 85097 916 1

'Letts' is a registered trademark
of Charles Letts & Co Ltd

Printed and bound in Great Britain by
Charles Letts (Scotland) Ltd

Acknowledgements:
The author and publishers are grateful to the
following for permission to reproduce photographs in
this book—

Ardea, London: *Jean-Paul Ferrero* 30; *J L Mason* 37;
Alan Weaving 81; *François Gohier* 92—British Dental
Association: 143—Electricity Council: 99—The
Environmental Picture Library: *Phil Brown* 40; 42;
V Miles 54; *Sue Cunningham* 62, 149; *Stan Gamester*
68; *D Ellison* 79; *C Westwood* 86; *C Jones* 92;
N Dickinson 116; *M Bond* 129; *Paul Glendell* 139—
Ford Motor Company: 69, 139—Leslie Garland
Picture Library: *Leslie Garland* 65, 139—Robert
Harding Picture Library: *T E Clark* 29; *David Hughes*
29; 33; 62; *Didier Barrault* 87; *Philips Electrical Ltd*
109; *G & P Corrigan* 121; *Jobmate 30* 124; *Ian
Griffiths* 142; 148; 149—The Hulton Picture Company:
24; 63; 78; 90; 117; 154—Impact Photos: *Alan Brooke*
87—Local Radio Workshop: 140—London Features
International: *Simon Fowler* 34—Network South-East:
58—Newslink Africa, London: 129—Ann Ronan Picture
Library: 62—Science Museum Library: 71; 134—
Science Photo Library: *Richard Hutchings* 35; *CNRI*
36; *Prof. Marcel Bessis* 36; *Dr Peter Moore* 40;
Philippe Plailly 57; *Alex Bartel* 66, 86-7; *Vaughn
Fleming* 75; *Science Source* 75; *Dr Arthur Lesk,
Laboratory of Molecular Biology* 75; *Gordon Garrard*
79; *Damien Lovegrove* 80; *Chris Priest/Mark Clarke*
80; *Martin Doarne* 81; *Matthew Shipp* 82; *Peter
Menzel* 83, 116, 147; *Simon Fraser* 84, 86; *Sinclair
Stammers* 86; *Julian Baum* 90; *Dr Jeremy Burgess* 94;
99; *Steve Percival* 105; *Adam Hart-Davis* 119;
Malcolm Fielding, BOC Group 140; *Dr Morley Read*
149; *John Howard* 150; *Charles Lightdale* 151; *NASA*
155; *Dr Fred Espenak* 158; *NAOA* 159; *Smithsonian
Institute* 160—Ziton Ltd: 116.

They are also grateful to the following for permission
to reproduce these extracts—

page 38: Highfield, Roger: 'Researchers open the way
to screening test-tube babies for genetic diseases',
Daily Telegraph, 19 April 1990: © Daily Telegraph
plc.

pages 126-8: Clover, Charles: 'A touch of the tropics
is coming our way', *Daily Telegraph*, 14 March 1988:
© Daily Telegraph plc.

Every effort has been made to trace all copyright
holders but, if any have been inadvertently
overlooked, the publishers will gladly receive
information enabling them to rectify any error or
omission in subsequent editions.

CONTENTS

PREFACE

The National Curriculum in science has produced considerable changes in science teaching from 5-16 with subsequent changes from 16-19. This book is designed to help you through key stage 3. For many students this will be between the ages of about 11 and 14 but may be all that some students require to prepare for new GCSE syllabuses.

This book covers all the attainment targets (ATs) set out for science in the National Curriculum (NC). In this book it has been considered that levels 3-7 correspond to key stage 3. By the end of this book you should have mastered all the topics required in science at key stage 3. The questions in each unit will help you check your understanding.

I would like to thank Graham Hill for his useful advice on the manuscript. Also I would like to thank the staff of Charles Letts; and Margaret Rawnsley for compiling the index. Finally, I would like to thank my wife, Judy, and my sons Robin and Timothy, for their help and support during the development of the project.

Bob McDuell

UNIT 1
INTRODUCTION

ABOUT THE NATIONAL CURRICULUM

As you complete the activities in this book and make progress through school, you will be following the National Curriculum. All pupils of your age across the country will be doing the same subjects.

The National Curriculum consists of ten subjects which you must study at school. These are divided into core and foundation subjects.

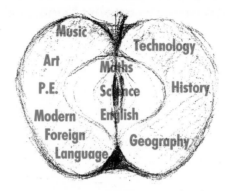

English, mathematics and science which help you in studying all the other subjects are the core subjects. The others are foundation subjects. Although it is not part of the National Curriculum, you will also study religious education at school.

Key stages

You are now at key stage 3 (which goes from age 11 to age 14). It is one of the four key stages which you go through as you complete your education to age 16.

Key stage 1: ages 5-7
Key stage 2: ages 7-11
Key stage 3: ages 11-14
Key stage 4: ages 14-16

Attainment targets

Each subject has its own objectives or goals which set out what you should be able to do at age 14. These are called attainment targets. Each attainment target has ten levels and you will progress through one level at a time. The average 14-year old will achieve level 5 or 6 depending on the subject although you might do better than this depending on how good you are at particular subjects.

Standard Assessment Tasks

When you are 14 and have completed key stage 3 you will be given a test, not set by your teacher, but common to all 14-year olds across the country. These tests (SATs) will help you to measure your progress through the National Curriculum.

Programme of study

Each subject also has a programme of study. This describes the work you have to do to meet the attainment targets at age 14. This book provides practice in the work that makes up the programme of study in science. By completing the activities in the book you will be much better prepared for the tests at age 14.

Science in the National Curriculum

At key stage 3 in science you are expected to make progress in seventeen attainment targets (AT). These are divided into two groups of related attainment targets called profile components, although profile component 1 in science has only one attainment target.

Profile Component 1: Exploration of science, communication and the application of knowledge and understanding

AT 1: Exploration of science

Profile Component 2: Knowledge and understanding of science, communication and the applications and implications of science

AT 2: Variety of life
AT 3: Processes of life
AT 4: Genetics and evolution
AT 5: Human influences on the Earth
AT 6: Types and uses of materials
*AT 7: Making new materials
*AT 8: Explaining how materials behave
AT 9: Earth and atmosphere
AT 10: Forces
AT 11: Electricity and magnetism
AT 12: The scientific aspects of information technology including microelectronics
AT 13: Energy
AT 14: Sound and music
AT 15: Using light and electromagnetic radiation
AT 16: The Earth in space
AT 17: The nature of science

The ATs marked * show topic areas which start at key stage 3.

Attainment target 1 is slightly different from all the rest. It is concerned with the ability to use and apply scientific skills as you study all the other attainment targets. So, whatever work you are doing, you will need to be able to: make predictions; carry out safe experiments; record findings in a variety of ways using ordinary and scientific language; draw conclusions; and make generalizations. If you are able to do these things as you develop your scientific knowledge then you will be learning to act and proceed like a scientist.

So, what kind of things will you be finding out as you cover key stage 3 science? You will learn about the rich variety of life on our planet and how different things survive, develop and reproduce. Materials of all kinds form a major part of our world and you will explore their types and uses, at the same time as explaining how they behave. You will begin to analyse and understand the chemical properties of materials and find out how new materials are made. Human beings are a major influence on the earth both for good and for bad and you will look at the different ways in which we affect the world around us. Considering the environmental implications of what we do will be a major part of this work. Energy forces, sound and light, to name but a few, will also be looked at in detail. You will also consider the place of the Earth in the wider universe;

a subject of much interest and fascination. Another attainment target is a particularly interesting one as it helps you to look at how scientific views and discoveries have changed over time. You will also begin to understand how people in different times and places first responded to ideas that we now accept as commonplace.

Don't be too worried if some of these subjects are unfamiliar to you at the moment. The National Curriculum is designed to develop your understanding as you progress through the different levels in each attainment target. This book will be of invaluable help to you as well.

Good luck and enjoy National Curriculum science!

HOW TO USE THIS BOOK

This book covers all the topics required in science at key stage 3.

Each unit covers a specific area of science and explains in simple language:

1 what you should know from key stage 2
2 what is required at key stage 3

Each of these units also gives you questions and activities to test your understanding. The answers to these questions can be found in the answers section which starts on page 166.

Scientific investigation is an important part of National Curriculum science and activities to develop this important skill will be found throughout the book in boxes labelled 'Action!' Many other investigations could be devised. Extra problems of this type can be found in the *Letts Science GCSE Coursework Companion*.

The boxes labelled 'Did you know?' explore the nature of science. They show how scientific ideas have developed over the centuries and how they both affect and are influenced by the society in which we live. You will find biographies of famous scientists and read about their great discoveries. You will also read about historical confusions, philosophical arguments and wrong turnings taken by scientists.

If you read a newspaper, or a scientific magazine, you will come across many other examples. You might do some research in a similar way about other scientists and their work. These could include:

Alexander Fleming   Robert Wilhelm Bunsen
Humphrey Davy   Alfred Wegener
Robert Boyle   Johannes Kepler

THE HISTORY OF SCIENCE

You may find it hard to get important scientific discoveries into the correct chronological order. The time chart on page 9 shows some of the important scientific advances and when they occurred. You will notice immediately that there has been more progress in the last two hundred years than in any previous time. This reflects the interest in the development of science. It is a fascinating fact that there are more scientists living today than the total number of scientists that have ever lived before.

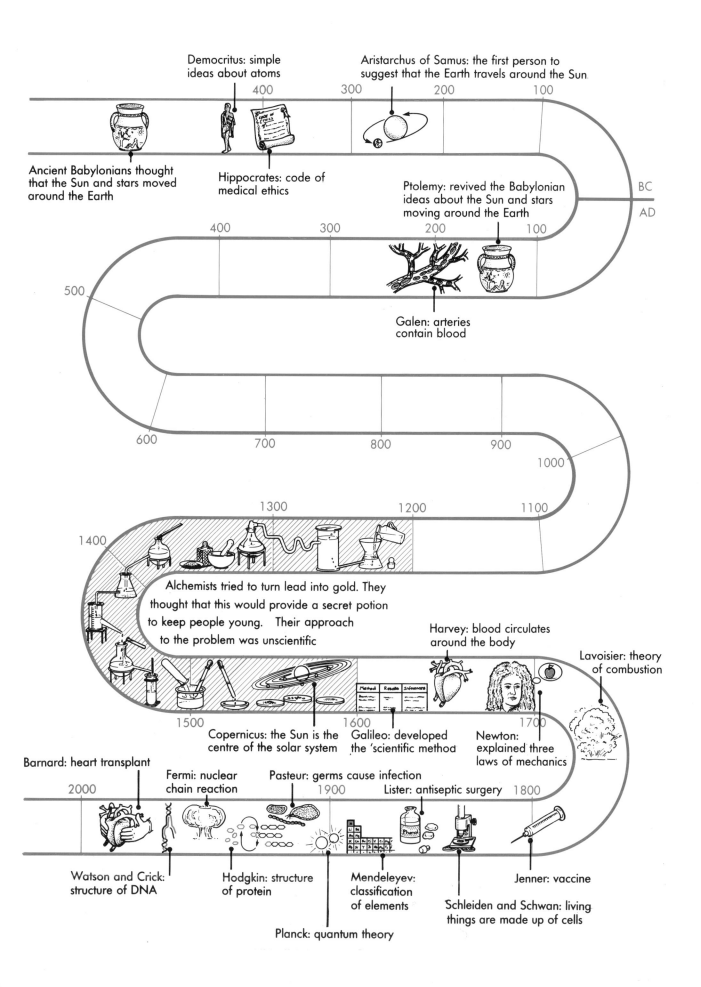

Democritus: simple ideas about atoms

Aristarchus of Samus: the first person to suggest that the Earth travels around the Sun

400 300 200 100

Ancient Babylonians thought that the Sun and stars moved around the Earth

Hippocrates: code of medical ethics

Ptolemy: revived the Babylonian ideas about the Sun and stars moving around the Earth

BC
AD

400 300 200 100

500

Galen: arteries contain blood

600 700 800 900 1000

1300 1200 1100

1400

Alchemists tried to turn lead into gold. They thought that this would provide a secret potion to keep people young. Their approach to the problem was unscientific

Harvey: blood circulates around the body

Lavoisier: theory of combustion

1500 1600 1700

Copernicus: the Sun is the centre of the solar system

Galileo: developed the 'scientific method'

Newton: explained three laws of mechanics

Barnard: heart transplant

Fermi: nuclear chain reaction

Pasteur: germs cause infection

Lister: antiseptic surgery

2000 1900 1800

Watson and Crick: structure of DNA

Hodgkin: structure of protein

Mendeleyev: classification of elements

Jenner: vaccine

Schleiden and Schwan: living things are made up of cells

Planck: quantum theory

9

ACTION!

Draw a time chart of some of the major scientific and technological inventions of the last four centuries. Your chart should include advances made between the years 1500 to 2000.

The list below will give you some starting ideas but you will need to look up the missing information from reference books. Include your own inventors on your chart. How many famous women can you find?

Date	Invention	Inventor
1590	compound microscope	Zacharias Janssen
1593	thermometer	
	telescope	Hans Lippershey
1643		Evangelista Torricelli
1650	air pump	Otto von Guericke
1714		Gabriel Fahrenheit
1800		Count Alessandro Volta
1822	camera	
	dynamo	Michael Faraday
	Bunsen burner	Robert Bunsen
1866		Alfred Nobel
1870	margarine	Hippolyte Mege-Mouries
1876	telephone	
1885	vacuum flask	
1892	zip fastener	Whitcomb Judson
1895	wireless	
	tape recorder	Valdemar Poulsen
1908	bakelite	Leo Baekeland
1913		Hans Geiger
	tungsten filament lamp	Irving Langmuir
1930	jet engine	
1935	nylon	Wallace Carothers
	electron microscope	Vladimir Zworykin
	electronic digital computer	J. Presper Eckert & John Mauchly
1948		Chester Carlson
1960		Theodore Maiman
1981	artificial heart	Dr Robert Jarvis

Table

You may be able to find some other inventions to add to the list.

Do you notice any pattern to the type of inventions that scientists have developed over the last few hundred years?

UNIT 2
THE VARIETY OF LIFE

You should already be aware of the wide variety of different living things. Plants and animals have particular needs if they are to remain alive. For example, water is essential for plants and animals. You will probably have experience of looking after plants or animals. This will mean that you have treated them with care and provided for their needs.

SPOTTING SIMILARITIES AND DIFFERENCES

This picture is the kind of 'spot the difference' puzzle which often appears in magazines.

Can you spot the ten differences between the two diagrams?

This kind of puzzle requires you to look closely at a detail in one of the pictures and then compare it with the same part of the other picture. Some differences are obvious but others take much more finding.

We often do similar investigations in science when we compare two similar living things.

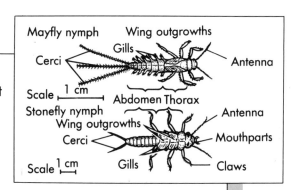

 ACTION!

Nymph spotting

Look at these two young forms of different insects—the stonefly and the mayfly. These immature forms are called nymphs. What similarities can you see between these two nymphs? The table lists differences between these two nymphs.

Stonefly nymph	Mayfly nymph
long antennae	short antennae
large body	small body
projecting mouthparts	no mouthparts visible
large head	small head
distinct neck	no neck
three visible segments in thorax	only two visible segments of thorax
wing outgrowth on two segments	only one pair of wing outgrowths
gills on thorax	gills on abdomen
pair of claws on each leg	single claw on each leg
no hairs on first part of first leg	hairs all along first leg
two cerci	three cerci
ten abdominal segments	eight abdominal segments

Now look for the differences between two other similar insects. Try to find at least five differences between the two creatures.

Spot the differences

(a) Look at the two specimens A and B and list three differences between them.

(b) Similarly, list four differences between C and D.

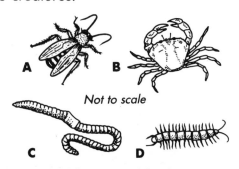

Not to scale

All creatures great and small

Biology is the study of living things. The word comes from a Greek word, bios, meaning 'life'. It is easy to distinguish a living thing from a non-living thing. For example, a living thing can grow and a non-living thing cannot. Can you think of any other differences?

There are millions of different **species** of plants and animals. New species are being discovered all of the time. Rather than try and study each one separately, it is sensible to put them together and study them as groups.

For example, if we consider the animal kingdom, we can divide all animals into two major groups: animals with backbones; and animals without backbones. Animals with backbones are called **vertebrates** and animals without backbones are called **invertebrates**.

ACTION!

Spine time

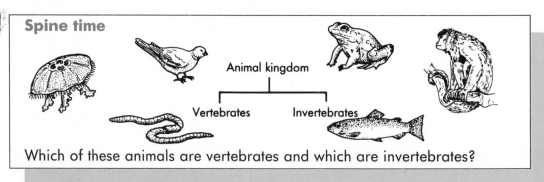

Which of these animals are vertebrates and which are invertebrates?

We can divide vertebrates further according to their body temperatures. An animal with a body temperature the same as its surroundings is called **cold blooded**. An animal, able to keep its body temperature the same even on hot or cold days, is called **warm blooded**.

ACTION!

In hot and cold blood

Which of these animals are warm blooded and which are cold blooded? Remember a cold-blooded animal cannot control its body temperature. Its body temperature is the same as the temperature of its surroundings.
Warm-blooded animals control their body temperature and keep it at a constant level.

12

The table below lists common groups of vertebrates divided into cold-blooded and warm-blooded **classes**.

Cold blooded	Warm blooded
fish	birds
amphibians	mammals
reptiles	

Each of these classes have there own characteristics.

Fish	Paired fins, gills
Amphibians	Slimy skin. Spend some of their lives in water
Reptiles	Dry scaly skin. Lay eggs on land
Birds	Feathers. Lay eggs on land
Mammals	Hair. Provide milk for young from special glands

ACTION!

Class creatures

Look at each of the illustrations and decide which one of these five classes each animal belongs.

Fish
Mammal
Amphibian
Reptile
Bird

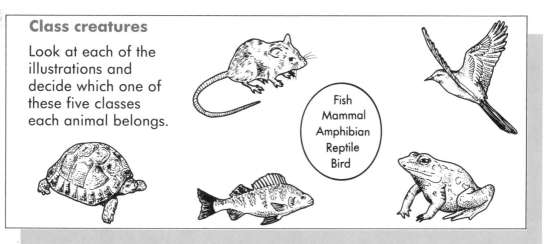

Did you know?

Carolus Linnaeus (1707-1778)

Carolus Linnaeus was born in Sweden in 1707.
He was the son of a church minister.
He went to medical school to train as a doctor,
but his main interest was in plants.
He spent much of his spare time searching the countryside for
new plants and then identifying them. He visited Lapland and wrote a book
about the plants there. He even visited Britain and fell in love with the yellow
gorse bushes of our heathlands. He is said to have fallen onto his knees in
thanksgiving at the sight!

Linnaeus realized that the same name was often given to more than one
plant. Sometimes a plant would even be given different names in different
parts of the same country. He felt that every plant or animal should have its
own individual scientific name which people throughout the world would
recognize.

He used the ideas of the English botanist John Ray, who had already started
using Latin to name plants. The naming system he devised consists of two
words. The first word gives the **genus** (or general name) which the
organism belongs to and the second word gives the **species**. For example,
the columbine which flowers in our gardens in May is called *Aquilegia
hybrida*. *Aquilegia* is the genus and *hybrida* is the species.

Marianne North was one of the most famous
women botanists of the 19th century.
Botany, the study of flowers and plants, has interested a number of scientists.

Marianne visited the Royal Botanical Gardens at Kew where she drew plants
and studied their behaviour. She travelled through Europe and the Middle
East, making detailed records of the plants she saw. After the death of her
father in 1869, she decided to visit other parts of the world to see plants
growing in their natural environments. She travelled to Jamaica, North and
South America, Japan, India, Australia, South Africa and the Seychelles.
Of course, travelling was not as quick and easy as it is today. As she
travelled, she discovered four new plant varieties and brought many new
plant species back to England. A special art gallery was built at Kew to
display her paintings.

She never married and used money left by her father on his death to support
her travels and work. Her married sister, Catherine, also painted flowers.
However, she did not have the opportunity to travel and her work today is
largely unknown.

Apart from ease of travel, what other advantages would Marianne North
have if she was carrying out her studies today?

We can classify invertebrates in a similar way producing a number of different
groups of **phyla** (singular-phylum).

Phylum	Feature
Protozoa	Made of one cell, e.g. amoeba
Sponges	Animals made of cells loosely joined together
Cnidaria	Body walls made of two layers of cells, e.g. jellyfish, sea anemones
Flat worms	Flattened worm-like shape
Annelida	Worms made of segments, e.g. earthworms
Arthropoda	Jointed legs, bodies made of segments includes spiders, insects, centipedes
Mollusca	No segments. A fleshy pad on which they crawl, e.g. slug, snail
Echinodermata	Star shaped pattern-spiny skin, e.g. starfish

The diagram at the top of page 15 summarizes the family tree of animals. The
number in each case is an estimate of the number of different species. In the
same way we can classify plants into different groups. The plant kingdom is
divided as follows:

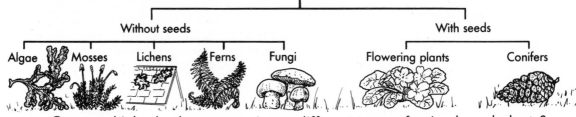

Can you think why there are so many different types of animals and plants?

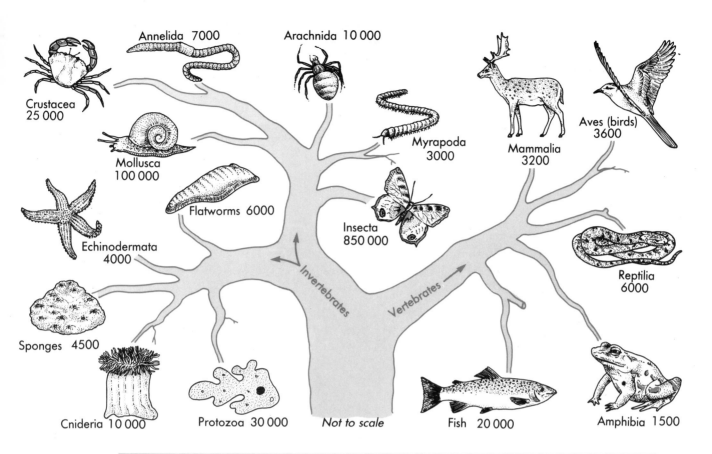

Crustacea 25 000

Annelida 7000

Arachnida 10 000

Mollusca 100 000

Myrapoda 3000

Mammalia 3200

Aves (birds) 3600

Flatworms 6000

Echinodermata 4000

Insecta 850 000

Invertebrates

Vertebrates

Reptilia 6000

Sponges 4500

Cnideria 10 000

Protozoa 30 000

Not to scale

Fish 20 000

Amphibia 1500

> **Did you know?**
>
> Do you realize that if you went out into the rain forest with a net looking for insects and spiders, there is a very good chance that you will find a species which has never been found before. It is believed that at least three-quarters of tiny creatures like insects and spiders have never been positively identified. It is possible that a new species could be found every day.

KEYS

Having seen the wide diversity and groupings of living things, we are going to use **keys** to identify some of them. A key is a series of questions which leads to the identification of individual plants or animals. At each point in the key, the question divides the organisms into two groups, depending upon whether or not they have a certain characteristic.

Look at the key below that identifies seven species of gull. The key can be expressed in another way as in the table at the top of page 16.

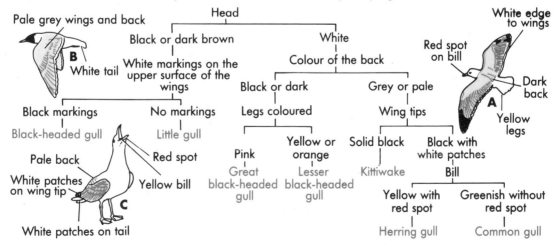

15

1	Head black or dark brown	Go to **2**
	Head white	Go to **3**
2	Black markings on upper surface of wings	Black-headed gull
	No black markings on upper surface of wings	Little gull
3	Back black or dark	Go to **4**
	Back grey or pale	Go to **5**
4	Legs flesh coloured	Great black-backed gull
	Legs yellow or orange	Lesser black-backed gull
5	Wing tips solid black	Kittiwake
	Wing tips black with white patches	Go to **6**
6	Thick yellow bill with red spot	Herring gull
	Thin greenish bill without red spot	Common gull

Keys use differences between plants, or animals, to help to identify them.

Use the key to identify the three gulls drawn in the key on page 15.

ACTION!

The key to leaves

The key can be used to identify six leaves. Use the key to identify the six leaves.

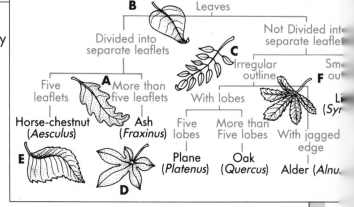

Name that fish

The drawing shows six fish. Try to make your own key to identify each fish.

You could start by asking questions such as: *Has the fish got a tail fin?*

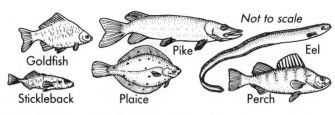

The eel has no tail fin. The other five fish have tail fins. Now carry on! There are many different ways of doing it.

Identify the creatures

The picture shows four arthropods found in a wood. Use the key to identify the four organisms.

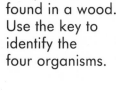

16

THE RHYTHMS OF LIFE

Living things respond to seasonal and daily changes in different ways.
Here are some examples of this:

1 A deciduous tree (e.g. oak) sheds its leaves in autumn and grows new leaves in spring. The pigments in the leaves change at different times of the year.

2 A tortoise is an example of an animal which hibernates during the winter. The tortoise's metabolism slows down during the winter enabling it to conserve energy and live through harsh conditions. It uses up a store of energy during hibernation. In spring the tortoise becomes active again.

3 A chrysanthemum comes into flower in the autumn when the length of the day is decreasing. Nurserymen, wanting to get chrysanthemums in flower at other times of the year, have to reduce artificially the amount of light each plant receives.

4 Nocturnal animals become active at night. In a zoo, nocturnal animals are kept in the dark so that visitors can see them moving about.

Can you think of any other examples?

HABITATS

A place where a community of organisms live is called a **habitat**. The drawing shows some plants and animals which live in a wood. This kind of environment is called a terrestrial habitat. Conditions in the wood are often cool and damp with limited amounts of light reaching the ground.

Not to scale

The picture on page 18 shows plants and animals which live in a pond. This is a very different environment from a wood, so different plants and animals will live there. A pond is an example of a aquatic habitat. Every organism lives in a habitat where it can survive successfully. How can we tell if a species is surviving well in a habitat? What factors affect the number of creatures that live in a habitat?

Not to scale

Some animals move away from light (e.g. woodlice, earthworms and cockroaches). This is because light is associated with warmth which would cause such animals to lose moisture, dry out and die. They tend to thrive in cool, damp conditions, (e.g. under stones).

Trees in a forest grow tall to reach the light which is necessary for making food by photosynthesis and for growth. Mosses and ferns are happier living in the shade of the trees in the forest. There it is cooler and there is less chance of drying out.

ACTION!

Finding a niche

The two nymphs on page 11 live in fast, flowing rivers. They crawl around underneath stones. Why are they suitable for survival in this habitat?

Surveying the sites

The table shows animals collected in two sites:
Site A, a damp, shaded area of the garden;
Site B, a dry open pasture with a sunny position.

Animal	Number of animals found at location	
	A	**B**
snails	55	1
worms	20	5
centipedes	5	1
ants	40	25
spiders	20	15
beetles	20	10
aphids	20	40

Not to scale

(a) Which animal occurred in larger numbers at site B rather than site A?

(b) Complete the pie chart opposite, showing the distribution of animals at site A.
Each segment is 20°.

(c) Which animal is likely to be a secondary consumer? Give a reason for your answer.

(d) Suggest **two** reasons why there are more snails at site A than at site B.

FOSSILS

Fossils are traces of extinct organisms preserved in rocks. Scientists expect such rocks to be older than a few ten thousand years to qualify as a fossil. Most fossils were formed when the remains of plants and animals were buried by sediment. Only the hard part of the organism will remain, (e.g. the shell or an impression of the shell in a rock). The scientist has to decide what the original creature was like by looking at similar organisms living today.

The types of fossil found in a particular area, give some idea of the organisms which existed in that location. The picture below shows some fossils found in the Peak District, in Derbyshire. All of these fossils are remains of sea creatures. Scientists therefore know that Derbyshire was under the sea when these animals were alive about 300 million years ago

Bivalves

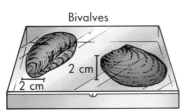

Brachiopods

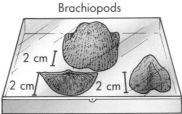

Corals

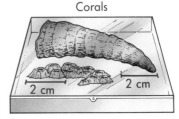

ACTION!

Making your own fossil

This investigation will help you to see what took place over millions of years. Take a shell and press it into a piece of plasticine. Make sure you have a clear imprint of the shell.

Now mix some plaster of Paris with water until you have a very thick, smooth paste. Pour the plaster into the plasticine mould and leave the plaster to set. Carefully remove the plasticine and you will have a plaster copy of the original shell. Fossils formed in the earth in a similar way with other rocks taking the role of the plaster.

How would you guess the age of a real fossil? Find out how scientists discover this information!

FEEDING RELATIONSHIPS

In a habitat, organisms are interrelated because of how they feed and how their dead remains decay. Organisms can be divided into groups of **producers** and **consumers**.

Producers are green plants which change light energy to trapped chemical energy by photosynthesis. Examples are grass (on land) and duckweed (in water).

Look at the four organisms below. The cabbages are producers. The slugs, feed on the cabbages and so are called **primary** consumers (or herbivores). The pigeons which feed on the slugs are called **secondary** consumers. The human being who feeds on the pigeon is the **tertiary** consumer.

This relationship can be summarized by a **food chain**.

cabbage → slug → pigeon → human

All food chains begin with a producer. The secondary and tertiary consumers that eat only animals are called carnivores. Those consumers that eat both animals and plants are called omnivores.

The slug in the example is the **prey** and the pigeon is the **predator**. However, the pigeon is the prey for the human.

The diet of an organism is more complicated than can be shown by a simple food chain. The complicated relationships between organisms can be shown by a **food web**.

Tertiary consumers — Fox Owl Huma

Secondary consumers — Frog Hedgehog Stoat Pig

Primary consumers — Snail Caterpillar Vole Rabbit S

Primary producers — Green plants, shrubs and trees

When plants and animals die, all of the nutrients stored in the body are recycled by **decomposers**. These organisms break down the dead animals and plants, and release nitrogen into the soil.

Decomposers include fungi and bacteria. They produce digestive juices which act on a dead plant or animal making it decay and become soluble.

ACTION!

Spinning a food web

Here is some information about plants and animals which live in a seashore environment:

(a) human beings eat crabs, prawns and mussels
(b) mussels eat tiny plants and tiny animals
(c) gulls eat starfish, mussels and crabs and prawns
(d) starfish and crabs eat mussels
(e) prawns eat tiny animals

Use this information to draw a food web.

Predator v prey

Look at the pictures and match a predator to its prey. For example: owl (predator) and small shrew (prey).

Not to scale

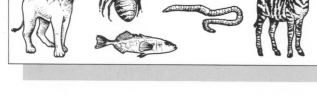

Primary consumers eat producers (e.g. slugs eat cabbage). Only about 10 per cent of the trapped energy in the cabbage becomes part of the slug's body. When the slugs are eaten by pigeons only 10 per cent of the energy in the food becomes part of the pigeon's body. The further up the chain, the less and less energy is passed along the chain. This leads to:

1 Fewer organisms being supported as the food chain is ascended. This is summarized opposite and is called the **pyramid of numbers**.

Tertiary consumers — Humans
Secondary consumers — Pigeons
Primary consumers — Slugs
Primary producers — Cabbages

2 Less total body mass of organisms (biomass) as the food chain is ascended.
The mass of cabbages is greater than the mass of slugs, which is greater than the mass of pigeons. This is called the **pyramid of biomass**.

Humans
Pigeons
Slugs
Cabbages

BIO-CYCLES

Nitrogen, carbon and water are recycled in nature.
When organisms die, or produce waste products, these are decomposed and the raw materials are released to be used again.

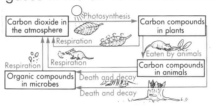

Look at the nitrogen cycle which summarizes the ways that nitrogen is added and removed from the soil.

When plants are full-grown they remove nutrients from the soil (e.g. nitrogen). If the land is to remain fertile, this nitrogen has to be replaced. It can be replaced naturally during thunderstorms and when plants and animals die and decay. Certain plants are able to absorb nitrogen directly from the air and fix it into the soil (e.g. clover, peas and beans).

In the form of nitrates, nitrogen can be washed out of the soil into rivers and lakes. High levels of nitrates in drinking water can cause health problems. Find out which problems this may cause!

The carbon cycle is summarized below. Photosynthesis by plants ensures that the percentages of various gases in the air remain constant. Can you think why?

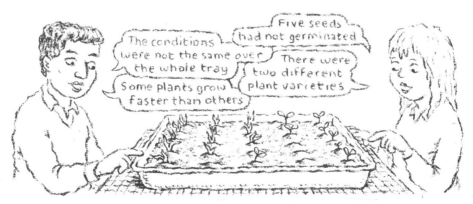

ACTION!

Growing plants from seeds

Majid and Sarah sowed 25 seeds in a seed tray using a compost. After two weeks they looked at the seedlings.

> The conditions were not the same over the whole tray
> Some plants grow faster than others
> Five seeds had not germinated
> There were two different plant varieties

They made four comments about their investigation. Some of their comments are **observations** and some are **inferences**.

An observation involves noticing changes which take place. These changes may involve looking, smelling, touching, hearing or sometimes even tasting. Observation requires first-hand experience. An inference is a conclusion drawn from a series of observations. Inferences can be made from observations recorded by other scientists.

Which of the comments made by Majid and Sarah are observations and which are inferences?

ACTION!

Mrs Budd is a keen gardener and is very proud of her roses. They win all of the prizes at the local village show. However, aphids are a problem, especially after a mild winter. She looks in her gardening book and finds two methods of controlling them:

1 Spray the roses with an insecticide 'Kill-all' which comes in an aerosol container. It kills all the harmful aphids but also kills the harmless ladybirds. Ladybirds are a natural predator for aphids.

2 Spray the roses with soap solution. This is an old-fashioned remedy for aphids. It is not very effective but it does not kill the ladybirds.

Discuss the advantages and disadvantages of these two methods of controlling the aphids. What could be the effects on the environment? Can you think of any other ways of controlling the number of aphids?

PROJECT

Zoos Visit a zoo and look at the wide range of animals. Complete a table, like the one below, where you have been given the first entry.

Name of animal	Scientific name	Where found
Lion	*Panthera leo*	Central and South Africa

Then, complete a map of the world showing where different animals live. You should also find out the food that each animal needs and the dangers it faces in the wild.

Some people regard keeping animals in zoos as cruel but others believe that, if animals were not kept and bred in captivity, some species may have died out. Through careful management, endangered species can be bred in zoos and released into the wild again. Write down your views.

Find out more about animals which are threatened with extinction in the wild which need to be preserved in zoos. There are some useful addresses on page 176.

UNIT 3
PROCESSES OF LIFE

You should already know that all living things are able to produce young to continue life on Earth. This is called reproduction and the organisms are said to reproduce. You should also know that food, exercise, rest, hygiene, safety and the proper and safe use of medicines are important for everyday human living.

CELLS: THE BUILDING BLOCKS OF LIFE

All living things are composed of different kinds of **cells**. The picture below shows typical plant and animal cells as seen through a microscope.

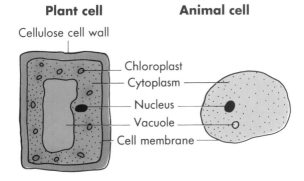

The differences between plant and animal cells are summarized in the table.

Plant cells	Animal cells
Chlorophyll present in chloroplasts	No chlorophyll or chloroplasts
Cellulose cell walls	No cellulose cell wall
Large vacuoles	Vacuoles small or absent

Human cells

The human body is made up from over 50 million tiny cells. Different cells have different purposes in our bodies. Apart from red blood cells, all cells have a **nucleus**. This controls the action of the cell. It also stores genetic information. The nucleus of the cell is surrounded by **cytoplasm** which, in turn, is surrounded by a **cell membrane**.

Cells are usually grouped together to form **tissues** where they work together. Examples of tissues in our bodies include: epidermis (skin); blood; bone; and nerves. Tissues themselves are grouped together to form **organs**. Examples of plant organs are: stems; leaves; roots; flowers; and seeds. The organs in a mammal (e.g. human) work together to form seven main systems in the body:

1 the **circulatory** system   **5** the **skeletal** system
2 the **respiratory** system   **6** the **nervous** system
3 the **digestive** system   **7** the **reproductive** system
4 the **excretory** system

The drawing on page 24 shows some of the major organs of the body.

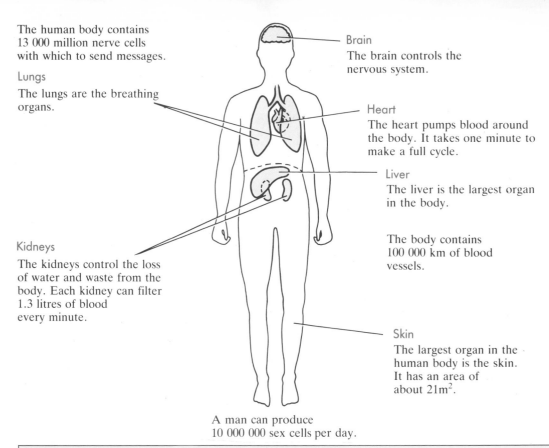

The human body contains 13 000 million nerve cells with which to send messages.

Brain
The brain controls the nervous system.

Lungs
The lungs are the breathing organs.

Heart
The heart pumps blood around the body. It takes one minute to make a full cycle.

Liver
The liver is the largest organ in the body.

The body contains 100 000 km of blood vessels.

Kidneys
The kidneys control the loss of water and waste from the body. Each kidney can filter 1.3 litres of blood every minute.

Skin
The largest organ in the human body is the skin. It has an area of about 21m^2.

A man can produce 10 000 000 sex cells per day.

Did you know?

Louis Pasteur (1822-1895)

In 1765, Lazzaro Spallanzani showed that food would not go bad if the microbes in it were killed. One method of killing the microbes in soup is to boil the soup.

Louis Pasteur was a French scientist who studied microbes, to stop food from going bad. The experiments he carried out in 1862 are shown below.

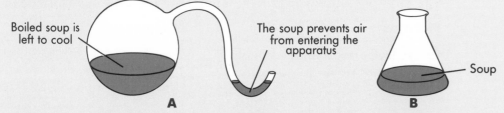

Boiled soup is left to cool

The soup prevents air from entering the apparatus

Soup

A

B

The soup in flask A does not go bad. The same soup in flask B goes bad.

Pasteur was able to isolate the germs which caused diseases such as cattle anthrax and chicken cholera.

He was one of the first people to carry out vaccinations to prevent illness.

How can you explain why the soup went bad in B but not in A?

ACTION!

Amoeba: a single-celled organism

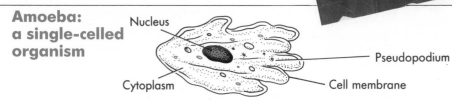

Nucleus

Pseudopodium

Cytoplasm

Cell membrane

The diagram on page 24 shows a single-celled organism called an amoeba. This is found in ditches and stagnant water.

Look at the diagrams below which show amoeba engaged in three activities. What is happening in A, B and C?

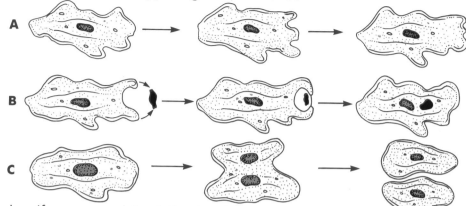

Jennifer proposed the following theory or **hypothesis**:

'Amoeba prefer to live in a dark environment rather than a light one'

How could she attempt to test this hypothesis? Why should she repeat this experiment a number of times?

THE RESPIRATORY SYSTEM

Respiration involves three processes in the human body:

1 oxygen is taken to the muscles by the blood

2 energy is released by the reaction of the food with oxygen

3 carbon dioxide is taken back to be breathed out into the air

There are two sorts of respiration: **aerobic respiration**; and **anaerobic respiration**.

Aerobic respiration requires oxygen. It produces more energy than anaerobic respiration as the food, the fuel, is completely burnt.

glucose + oxygen → carbon dioxide + water + energy

Anaerobic respiration takes place without oxygen and releases less oxygen. It takes place in muscles when there is less oxygen, (e.g. in a strenuous race). Lactic acid is produced.

glucose → lactic acid + energy

In a 100 m race, an athlete builds up about 40 g of lactic acid because oxygen is used up faster than the body can take it in.

At the end of the race the athlete has to breathe deeply to replace the oxygen used up.

A marathon runner does not build up an oxygen debt but uses up oxygen at the rate at which it is taken into the body.

The diagram shows the human respiratory system.

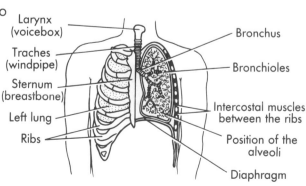

Larynx (voicebox)
Traches (windpipe)
Sternum (breastbone)
Left lung
Ribs
Bronchus
Bronchioles
Intercostal muscles between the ribs
Position of the alveoli
Diaphragm

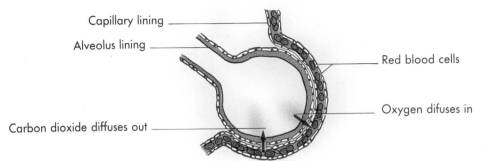

Capillary lining

Alveolus lining

Red blood cells

Oxygen difuses in

Carbon dioxide diffuses out

Air enters the lungs which consist of branched tubes, ending in millions of tiny sacs called **alveoli**.

The walls of the alveoli are extremely thin and there is a very large surface area within them. Oxygen can diffuse through the alveoli into the blood from the heart, and carbon dioxide can diffuse from the blood into the alveoli. Rich in oxygen, the blood is then returned to the heart. The blood transports oxygen to the muscles and carbon dioxide back from the muscles.

Exercise makes us breathe more quickly so that more oxygen is drawn into the lungs. Smoking, air pollution and diseases like bronchitis, can affect the working of the respiratory system. Find out how!

THE DIGESTIVE SYSTEM

Digestion is the breaking down of large insoluble food molecules into small molecules which the body can then use. The digestive process involves:

1 a mechanical breaking down of the food, (e.g. by chewing)

2 a chemical breaking down using enzymes and acid. Enzymes are biological agents which speed up reactions

The main chemicals (nutrients) in food are proteins, carbohydrates and fats.

The table summarizes the chemical changes which take place:

Food	Location of Enzymes	Products
proteins	stomach, pancreas, small intestine	amino acids
carbohydrates	mouth, pancreas, small intestine	sugars
fats	pancreas, small intestine	glycerol, acids

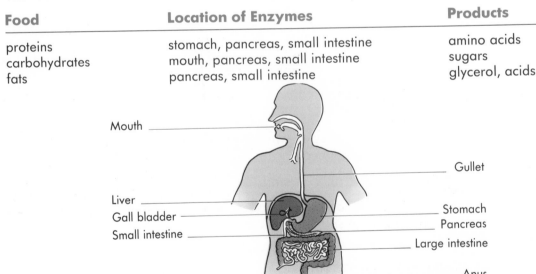

Mouth

Gullet

Liver

Gall bladder

Small intestine

Stomach

Pancreas

Large intestine

Anus

The picture shows the digestive system. Digestion begins in the mouth where mechanical breaking down of the food occurs. The food is mixed with **saliva**–which contains the enzyme amylase. This starts to digest the large starch molecules. The food is swallowed and reaches the stomach. The stomach contains hydrochloric acid and enzymes which can break down large protein molecules.

The partly digested food is then passed into the small intestine where enzymes from the pancreas continue the digestion. The small molecules produced include:

glucose; and amino acids. These small molecules are absorbed through the lining of the gut into the bloodstream. The blood transports these small molecules around the body. Undigested food is passed into the large intestine where water is absorbed and faeces are formed. The faeces pass out of the body through the rectum and anus. Find out the kinds of chemicals that the human body cannot digest.

You are what you eat!

For good health a balanced diet is essential. The amount of the different types of food that people require varies with age, occupation and life-style. The main types of food are shown in table 1.

Type of food	Benefit to the body	Source
carbohydrates	provide energy, (e.g. sugar, starch)	bread, potatoes
proteins	provide amino acids for building and repairing the body	meat, fish, milk, cheese
fats	store energy	butter, oil
vitamins	required in small amounts for good health, (e.g. vitamin C)	fruit, vegetables
minerals	required in small amounts for good health, (e.g. iron)	fruit, green vegetables

Not to scale

In addition to the types of food in table 1, water and fibre (roughage) are also required. Fibre is not digested but helps in the production of faeces and prevents constipation. There is also evidence that fibre in the diet reduces risks of bowel cancer.

ACTION!

Food for thought

Table 2 gives the composition of two different brands of fruit yoghurt:

Ingredients per 100 g	Brand X	Brand Y
energy value	150 kJ	480 kJ
protein	4.5 g	4.0 g
fat	0.3 g	2.5 g
carbohydrate	5.2 g	18.4 g
additives	preservative, artificial sweetener	preservative

(a) Which brand of yoghurt would be more suitable as part of a slimmer's diet? Explain your reasoning.

(b) Brand Y contains no artificial sweetener. What do you think sweetens this brand of yoghurt?

(c) What method of storage would you recommend if these yoghurts did not contain preservative?

(d) Calculate the mass of protein, fat and carbohydrate in one 125 g tub of brand Y.

Food Additives Food manufacturers put additives in food for various reasons. These include:
1 making it look good
2 keeping it from going bad
3 keeping the ingredients thoroughly mixed
4 reducing the costs of ingredients
5 improving the flavour

Look at food labels on different products to find out their ingredients. For example:

CHERRYADE
Ingredients: Water; sugar; citric acid; flavouring; artificial sweetener (sodium saccharin); preservative (E211); colours (E122, E124).

CHEESE SPREAD
Ingredients: Cheese; skimmed milk powder; butter; whey powder; emulsifying agent (E339); preservative (E202); colour (E160).

Food additives approved for use in the European Community (EC) are given **E numbers**, for example:

E100–E180 are colouring agents Many colouring agents are artificial, sometimes even made from coal! Some natural colourings are used (e.g. caramel). You cannot assume that a natural colouring is healthier than an artificial one.

E200–E290 are preservatives They stop microbes growing in the food.

E300–E321 are anti-oxidants They stop air turning the food bad.

E322–E494 are emulsifiers These ensure that the ingredients remain mixed.

Apart from these additives there are sweeteners, flavourings and flavouring enhancers.

An orange squash is advertized as containing no artificial sweeteners. Could sugar be used in the manufacture of this squash?

Food scares?

People are getting very concerned about food additives. Some have been shown to be harmful causing medical problems including allergies.

Do a survey in your local supermarket to find out foods which contain additives?

Take care to plan your investigation carefully.

You could ask people questions about whether they look at labels on food or whether they avoid certain additives, etc.

Your survey could lead to some interesting discussions. It is certainly a subject which could affect your health!

REPRODUCTIVE SYSTEMS

Reproduction is the making of new organisms similar to the parent or parents. There are two main types of reproduction: **asexual**; and **sexual**.

Asexual reproduction

Asexual reproduction involves only one parent. It produces offspring which are identical to the parent. Examples of asexual reproduction include:

1 Single cell organisms such as amoeba and bacteria. They reproduce by growing and splitting into two identical halves.

2 Part of a plant can grow into identical plants. Stem cuttings, leaf cuttings, etc. will root and produce plants which are all identical.

As the offspring are identical to the parent, variation does not occur with asexual reproduction. Also any disease, or fault, in the parent will be present in the offspring. Can you see why?

Sexual reproduction

Sexual reproduction usually involves two parents. Both the male and the female produce sex cells. A human male sex cell (or sperm), shown below, is produced in his testes. The human female produces a female sex cell (or egg) in her ovary.

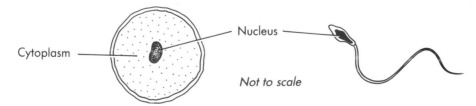

Cytoplasm — Nucleus —

Not to scale

The diagrams on page 30 show the main parts of the male and female reproductive systems.

29

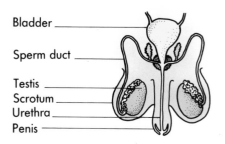

Bladder

Sperm duct

Testis
Scrotum
Urethra
Penis

Fallopian tube
Ovary
Uterus (womb)

Cervix

Vagina

During sexual intercourse sperm passes from the male's body through the penis into the female's vagina. The sperm enters the uterus and travels along the egg tubes. If there is an egg in the tubes, fertilization may take place. The fertilized egg travels to the uterus and beds itself into the lining. Here the embryo develops and, after nine months of pregnancy, the baby is born through the vagina.

A human baby is helpless at birth and relies on care from parents if it is to survive. Other mammals tend their young in a similar way (*see* photograph).

Some animals (e.g. fish) produce babies which do not need a great deal of care from parents. Can you explain why these animals need to produce lots of offspring?

If fertilization does not take place, the uterus lining breaks down and is lost as part of the monthly menstrual cycle.

Identical twins are formed if the fertilized egg divides into two parts and each part develops into a baby. They have the same genes and are either both male or both female. Non-identical twins are formed when two eggs are released from the ovary and both are fertilized. They do not have the same genes and can be both male, both female or one male and one female.

The structure of a flower

Inside a flower there are both male and female parts. These enable the plant to reproduce. The male parts are called **stamens** and they make pollen grains inside four-chambered **anthers**. The female parts, called **carpels**, are found at the centre of the flower. They consist of a hollow **ovary** which contains an **ovule**. When this receives pollen the flower starts to form a seed.

Many flowers have brightly coloured **petals** surrounding the carpels and stamens. Sometimes nectar, a sugary liquid, is produced at the **nectary** at the base of the petals.
Underneath the petals there is usually an outer ring of green **sepals** which protect the flower when it is still just a bud.

Male and female sex cells join together to fertilize the ovule.

This then develops into a seed.

The drawing summarizes the main parts of a flower.

Petal

Anther
Ovary
Nectary
Sepal

30

ACTION!

Parts of a buttercup

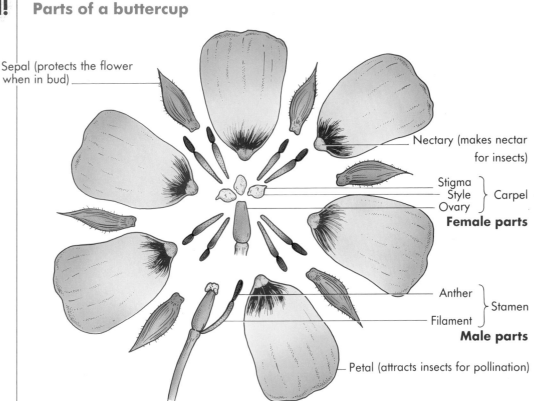

Sepal (protects the flower when in bud)

Nectary (makes nectar for insects)

Stigma ⎫
Style ⎬ Carpel
Ovary ⎭
Female parts

Anther ⎫
⎬ Stamen
Filament ⎭
Male parts

Petal (attracts insects for pollination)

The picture shows the various parts produced when a single buttercup flower is cut up. Use the diagram to complete the table by counting up the numbers of each type of plant.

Sepals	Petals	Stamens	Carpels

Now try the same exercise with other flowers. You will have to cut the flower carefully and lay out the pieces. It is a good idea to use sticky tape to stick each piece down so you do not lose any. Do you notice any patterns?
Can you explain your discoveries?

Flower facts

(a) Flowers are usually colourful, often scented and sometimes contain sweet nectar. Why do you think this is so?

(b) In a greenhouse in summer, a gardener is sometimes seen with a fine paintbrush, brushing inside first one tomato flower and then another. Why would the gardener be doing this?

UNIT 4
GENETICS AND EVOLUTION—
WHAT MAKES US DIFFERENT?

You should already know that individual human beings differ, and be able to measure some simple differences, (e.g. height, shoe size etc).
Some living things no longer exist because they have died out a long time ago. Dinosaurs and dodos are examples of animals that are now extinct.

EVERY KIND OF PEOPLE

Apart from red blood cells, all the cells in our bodies contain nuclei. These nuclei contain **chromosomes** which store information to enable new similar cells to be produced and information which can be passed on to future generations.
The genetic information is stored within chromosomes, on threads of **DNA**.
Every human cell contains 23 pairs of chromosomes. Twenty-two pairs can be matched up and are called **autosomes**. The other pair are the sex chromosomes (X and Y) which determine the sex of the person. If the sex chromosomes are alike (XX) the sex is male, and if they are different (XY) the sex is female.   *Female*

Sexual reproduction involves special cells called **sex cells**. When cells divide, during growth or repair, they produce identical cells. This process is called **mitosis**. After the division each cell contains the same number of chromosomes as the parent cell.

The diagram shows mitosis taking place in a cell with only four chromosomes.

The rest of the cell divides to form two new, identical cells

Two pairs of chromosomes

The chromosomes double

The chromosomes separate into two groups

The nucleus divides into two new nuclei, each with four chromsomes

Sex cells are not produced by mitosis but by a process called **meiosis**.
This time the chromosomes make an exact copy of themselves but the parent cell divides into four new cells. Each new cell has half of the number of chromosomes of the original cell.

Look at how this happens, again with a cell containing four chromosomes.

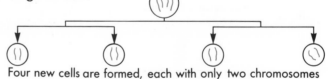

Four new cells are formed, each with only two chromosomes

The male and female sex cells join together during **fertilization** to produce a new cell, the **zygote**, which has characteristics of both parents and develops into the **embryo**.

This is summarized in the diagram.

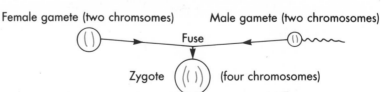

Female gamete (two chromsomes)   Male gamete (two chromosomes)

Fuse

Zygote   (four chromosomes)

Genes

The instructions carried by a chromosome for a particular characteristic, such as eye colour or blood group, are called **genes**. Every human being has two copies

of each gene in every normal body cell; one in each chromosome.
One gene comes from the father and one from the mother.

VARIATION

Have you noticed that brothers and sisters in the same family are not identical?
Unless there are identical twins, the chances of parents having two identical
children are about one in 1 800 000 000 000 000. This number is very large.
It is much larger than the number of people who have ever lived.
You will not meet your exact double during your lifetime.

Variation within a family is caused by new genes formed by mutation and
different mixes of genes. There are two types of variation:

1 Discontinuous variation This enables us to separate the population
into different clearly distinguished groups, (e.g. blood groups). We can sort the
blood groups of individuals into four main groups: A; B; AB; and O. No one falls
between two groups, (e.g. a mixture of groups A and O).

ACTION!

All about blood

The human body contains about five litres of a complex mixture called
blood. This is made in the marrow inside the bones. It contains:
red blood cells, which carry oxygen around the body;
white blood cells, which fight germs which get into the body;
platelets, which are tiny fragments which help the blood to clot and stop
bleeding; and
plasma, which is a colourless liquid in which the other cells exist.

(a) Label the
main parts of
the blood.

(b) In 1890 it was discovered that people could have one of four possible
blood groups. These are A, B, AB and O. If somebody needs blood
after an accident, it is important that they get the right type of blood.
The table below gives the percentage of people in different parts of
the world with each type of blood group.

Country	Percentage of people with each type of blood group			
	A	**B**	**AB**	**O**
England	42	8	3	47
Scotland	35	11	3	51
Ireland	32	11	3	54
China	23	26	6	45
Russia	36	23	8	33
Japan	36	22	9	32

The blood group of each person is inherited from their parents.
  (i) Which is the commonest blood group in each country?
 (ii) Which country has the lowest percentage of blood group B?
(iii) Why do you think that Russian people are more closely related to
      Japanese people than to Chinese people?

Another example of discontinuous variation is albinism.

This is a complete lack of skin pigment caused by the difference of a single gene.

An albino will never become tanned however much sun they are exposed to.

2 Continuous variation   Sometimes
we cannot see clearly different groups.

For example, if we were to measure the length of the middle finger of the right hand of thirty children, the results could not be clearly put into groups. Height and weight are other good examples. The graph shows the kind of variation which could be seen in the heights of a sample of men.

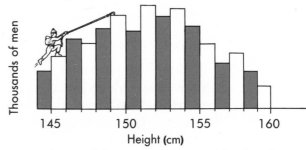

Whether this variation is due to genes inherited from parents, or whether it depends upon the way the men have lived, is the sort of question which has interested people for many years.

Scientists have studied identical twins to investigate this. Identical twins have exactly the same genes because they are both formed from a single fertilized egg which splits into two, developing into two identical embryos. Any differences between identical twins cannot be due to differences in their genes. Any differences must be due to the influence of their environment, such as amount and quality of food. Studying identical twins who are separated shortly after birth and reared separately can give interesting information. If the twins have similar characteristics, it suggests that inheritance was the main influence.

ACTION!

The differences between twins

A study was carried out on a hundred pairs of identical twins. For fifty pairs of twins, each pair was brought up together. For the other fifty, the pairs were separated at birth and brought up apart. The results are summarized in the table below:

Difference in characteristic	Twins brought up together	Twins brought up apart
height (cm)	1.6	1.7
weight (kg)	2.0	4.8

Look at the information in the table. Compare the differences in heights of twins brought up together and separately. Is there much difference?

Now, compare the differences in weights of twins brought up together and separately. Is there much difference this time?

Can you draw any conclusions from this information?

ACTION! Beetle features

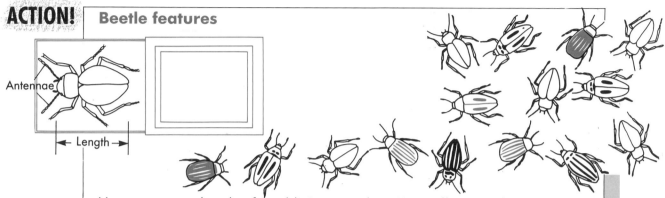

Antennae

←— Length —→

Here are some beetles found living together. You will notice they are not all the same. In the box you will see how we could measure the length of a beetle.

(a) What features do all the beetles have in common?

(b) What features about the beetles vary?

(c) Which features show continuous variation and which discontinuous variation?

(d) Display your results tables, bar charts or other ways which you think will show the results more clearly.

MUTATIONS

The copying of chromosomes when cells divide is very complicated. Mistakes can occur which are called **mutations**. Mutations can be caused by radiation and by some chemicals, agents called mutagens. These mutagens are found in certain drugs, cigarette smoke and fumes from certain types of plastics when burning.

Down's syndrome is caused by a mutation. Children with Down's syndrome have an extra chromosome. This occurs most frequently when the mother is old, for having children, and cell division to produce eggs has not occured properly.

35

INHERITED DISEASES

Haemophilia

Some diseases are passed on through genes and are inherited. For example, haemophilia is a genetic disease that stops blood clotting. It is caused by a mutant gene on the X chromosome. The effects of a single haemophilia gene in a male and female are different:

male The gene causing haemophilia is present and male suffers from haemophilia.

female The female does not show haemophilia because, although the haemophilia gene is present, it is masked by another more active gene. However, the female is a **carrier** of haemophilia and the condition may be seen in her children.

Phenylketonuria (PKU)

PKU is a genetic disease which affects one baby in 10 000. Babies with PKU suffer with brain damage soon after birth due to a high level of a particular amino acid. The enzyme which would break down this amino acid is not present because faulty genes prevent its formation. Nowadays, a blood sample at birth identifies babies suffering from PKU and a very restricted diet, containing none of the amino acid, prevents brain damage.

Sickle-cell anaemia

Sickle-cell anaemia is another inherited disease. Red blood cells contain haemoglobin. The haemoglobin carries oxygen around the body. Look at the red blood cells from:
(a) a healthy person;
(b) a person suffering from sickle-cell anaemia.

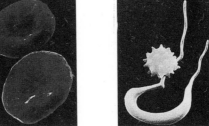

Notice in (b) that the blood cells have collapsed and formed an S shape. The person suffering from sickle-cell anaemia inherited the disease from his or her parents.

A baby inherits one chromosome from its father and one from its mother. On each chromosome there is a gene which determines whether red blood cells are round or sickle-shaped.

Key
D = Disc-shaped gene
S = Sickle-shaped gene

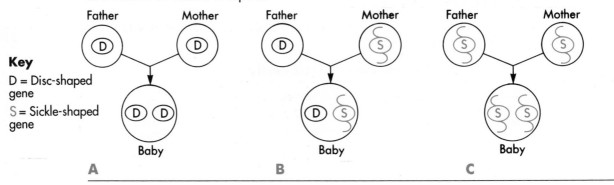

A

Both chromosomes contain genes producing circular cells. The baby produces cells.

B

One chromosome contains gene producing circular cells. The other produces sickle-shaped cells. The baby is healthy under normal conditions.

C

Both chromosomes contain genes producing sickle-cells. The baby suffers from anaemia because oxygen cannot be transported around the body. The baby may die.

Babies born with one chromosome containing one sickle-cell gene and one round-cell gene have one advantage. They cannot catch malaria. Can you think of a situation where such children may be at an advantage to babies without any sickle-cell genes?

NATURAL SELECTION: THE ORIGIN OF SPECIES

Charles Darwin proposed the idea of natural selection after observing many different plants and animals. One form of a gene may give the organism an advantage over the organisms with the other form of the gene (*see* sickle-cell anaemia)

An example of this is shown by the peppered moth. The peppering effect provides good camouflage for the moth. Moths are eaten by birds. Light-coloured moths on light backgrounds have a better chance of survival than dark moths on light backgrounds. The light-coloured moths survive to breed and so light moths predominate. The peppered moth has a speckled grey colour. Before 1850, all peppered moths were a light-grey colour. The black form was very rare. By the end of the 19th century, 95 per cent of all of the moths in industrial cities were black but in country areas the grey moths were still common (*see* 'Action!' (a)).

ACTION!

A peppering of moths

(a) Can you suggest why this happened? If you were to investigate the numbers of each type of moth living today, what would you expect to find?

(b) Light-coloured moths have an advantage over dark-coloured moths when viewed against light backgrounds. What difference would you expect if the moths lived where buildings were blackened by air pollution?

A good likeness

The table below shows information about two organisms which are similar in appearance:

	Honey-bee	Hoverfly
Body	3 segments	3 segments
Legs	3 pairs	3 pairs
Wings	2 pairs	1 pair
Colour	yellow and black stripes	yellow and black stripes
Length	1.5 cm	2 cm
Sting	yes	no

(a) Which of the organisms is larger?

(b) Apart from size, give one difference between the two.

(c) A predator will not eat either of these organisms, even though the hoverfly is harmless.
  (i) Explain the reason for the predator's behaviour.
  (ii) Explain how genetic mutations in the ancestors of the hoverfly account for the similarity between the two organisms.

PROJECT

1 Getting the gist Read the following article and then summarize the main points in the extract, in a maximum of 200 words.

Researchers open the way to screening test-tube babies for genetic diseases

By Roger Highfield
Science Editor

RESEARCHERS have discovered how to identify the sex of a test-tube baby before the embryo is transferred to its mother. The discovery by a team at the Hammersmith Hospital, west London, opens the way to screening for 200 genetic diseases.

Parents who carry disorders that would affect only boys can now ensure that only female test-tube babies are transferred to the mother. Three out of five women who took part in trials at the hospital are now pregnant after tests showed that their unborn children were female and normal.

The 200 diseases are all sex-linked. The most common is muscular dystrophy, which affects one in 3,000 live births.

The important and controversial development was achieved by Prof Robert Winston, Dr Alan Handyside, Miss Eleni Kontogianni and Miss Kate Hardy of the Royal Postgraduate Medical School at Hammersmith.

The three mothers-to-be are:
☐ Mrs Debbie Edwards, 29, of Hayes, Middx, who is expecting twins. She is a carrier of adrenoleukodystrophy, a progressive disease that starts out with loss of coordination and speech, leads to mental impairment and blindness, and is normally fatal.
☐ Mrs Christine Munday, 35, of Frimley, Surrey, also expecting twins, who

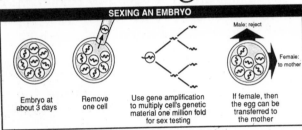

SEXING AN EMBRYO

Embryo at about 3 days — Remove one cell — Use gene amplification to multiply cell's genetic material one million fold for sex testing — Male: reject / Female: to mother. If female, then the egg can be transferred to the mother

already has a 12-year-old son suffering from a condition which causes mental and physical retardation.
☐ A woman, who wishes to remain anonymous, who is a carrier for Duchenne Muscular Dystrophy. She has had four unsuccessful pregnancies, one miscarriage and three terminations of affected pregnancies.

Prof Winston said: "For the first time, people who are carrying a severe genetic disease can start a pregnancy from the beginning, knowing it is normal, and do not have the spectre of a termination of pregnancy."

The announcement comes as MPs are preparing to vote next week on whether to allow research on embryos up to 14 days after conception. If embryo research had been banned, the Hammersmith work could not have taken place.

Prof Winston said: "There is no doubt

that this kind of work provides information about human embryology that will be of great benefit in the next two or three decades."

If the research continues, Prof Winston estimates his team could treat tens of patients this year, "perhaps even 100".

Prof Winston's team has also made important progress in extending the technique from checking the sex of the embryo to deciding — using a single cell — if it carries a defective gene responsible for a hereditary disease.

At present, genetic assessment of embryos is carried out after the embryo has become implanted in the wall of the mother's uterus, by sampling the amniotic fluid or by examining embryonic cells in the placenta in a process that can be done as early as eight weeks into pregnancy.

But the Hammersmith team has managed to determine the sex of embryos very much earlier, long before implantation, when the embryo consists of a ball of six or eight cells and is only a tenth of a millimetre across.

Dr Handyside, senior lecturer at the Hammersmith, was described by Prof Winston as offering the "science and real intelligence behind this".

In the procedure, the researchers take eggs from the mother and sperm from the father and fertilize the eggs in the laboratory. After almost three days, one of the cells is removed and studied using a technique developed by Cetus of California.

This allows the team to amplify the trace of genetic material contained in the single removed cell into sufficient quantities to test.

Prof Winston said another key development had been research to check that taking a cell from the developing embryo had caused no damage.

Several members of Mrs Edwards's family have had adrenoleukodystrophy. A nephew was healthy and normal until eight years old. Now, at 11, he is "blind, totally disabled, needs feeding and is incontinent and totally dependent on his family", she said.

Mrs Munday said her son was "very demanding" though "we love him dearly". She has tried for nine years to have a normal child.

Find out more about this issue by reading other articles in newspapers and magazines, (e.g. *Nature, New Scientist*). Research out the disadvantages of this scientific technique. Write a balanced account of the issues involved in genetic screening. Make sure that you include your own point of view in the conclusion to your report.

Think of other areas of scientific controversy, (e.g. nuclear energy, seal culling). By using a library and by contacting relevant organizations, study articles which focus on an issue taking care to read about both sides of the scientific argument. Some useful addresses can be found on page 176. Write a report as before. You can start a scrapbook of your views on controversial scientific issues by collating your reports with photographic cuttings from newspapers or magazines.

2 Seed Catalogues These can be interesting sources of information about different flowers and vegetables that can be grown from seed, (e.g. *Unwins, Histons, Suttons*).

Look in a seed catalogue and find out what is meant by the following terms: hardy annual; half-hardy annual; biennial; perenial.

Sometimes seeds are sold that are F_1 hybrids or F_2 hybrids. Find out the meaning of these terms. What are the advantages and disadvantages of F_1 hybrid seeds?

UNIT 5
HUMAN INFLUENCES ON THE EARTH

You should already be aware that human beings produce a wide range of waste materials. Some of these decay naturally but others are slow to decay.

THE EARTH UNDER THREAT

People have certain basic needs to live. They include food, air, water, shelter and warmth. In seeking to meet these needs for everybody, the Earth's valuable resources are used up producing certain waste products which can threaten the future of the planet.

Human needs are summarized below.

NEEDS

Food	We need food to provide energy and to build up our bodies. We rear animals to provide food.
Air	We need oxygen from the air to breathe.
Water	We need water for drinking and washing.
Shelter	We need some form of shelter usually made of available local materials.
Warmth	We need to keep warm. So we burn fuels and wear clothes. We also use fuels to cook our food.

People have used the Earth's resources for thousands of years. In the past however there was a large supply of resources and it was never necessary to consider problems which might be caused by shortages. Unfortunately, the resources are not shared equally around the world.

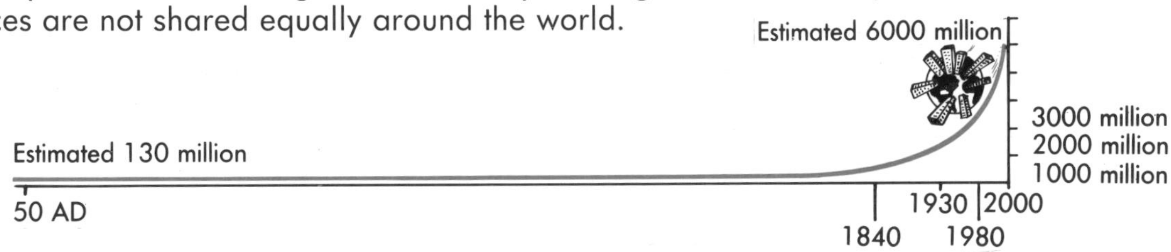

Estimated 6000 million

Estimated 130 million

3000 million
2000 million
1000 million

50 AD

1840 1930 2000
1980

The graph shows the growth of world population. Improving medical care and better conditions have enabled people to live longer, contributing to the growth in population.

By striving to provide all the resources that people need, the Earth's appearance has often suffered. Mining, to provide necessary minerals, produces scars on the landscape.

Destruction of forests to provide wood, and make more farm land, has an effect on the planet.

Factories producing vital goods also pollute air and water.

A load of rubbish

Every year in Great Britain, about 30 million tonnes of refuse are thrown away. Much of this refuse is tipped into holes in the ground or is burnt. It still contains valuable materials which could be salvaged and re-used.

We can divide materials which we throw away into two types: **biodegradable**; and **non-biodegradable**. Biodegradable materials, such as paper and vegetable waste, will rot away due to the effects of bacteria. Non-biodegradable materials, such as plastics and glass, do not rot away and remain largely unchanged. Therefore on a rubbish tip, some materials will rot away and others will not. The decay of biodegradable materials produces the gas methane which can build up in the tip and cause explosions.

Try to explain why these explosions happen!

The diagram shows the typical contents of our dustbins. About 90 per cent of this could be re-used again.

The re-use of materials is called **re-cycling**.

Kitchen waste 30%
Paper and card 25%
Dust 10%
Clothing 10%
Metal 8%
Glass 10%
Plastic 7%

ACTION! | **Drawing bar charts**

Use the information above to complete a copy of the bar chart on page 41.

40

The drawing below shows materials found in refuse which could be recycled.

Re-cycling saves using new raw materials from the Earth and so prolongs the life of limited supplies. It often saves energy costs too. For example, using re-cycled crushed glass in the glass-making process, rather than to make completely new glass, saves energy and is thus cheaper. It also provides additional employment in re-cycling industries.

AIR POLLUTION

Apart from the gases normally found in the air, other gases such as sulphur dioxide, oxides of nitrogen and carbon monoxide can be present. These are found because of human activity, especially burning fuels in homes, factories and motor cars. These gases cause **air pollution** and are called **pollutants**.

Sulphur dioxide is the major cause of air pollution. Coal contains about 2 per cent sulphur and when this is burnt it produces sulphur dioxide. In the atmosphere, sulphur dioxide reacts with water and eventually produces sulphuric acid. The term **acid rain** refers to rain that becomes appreciably acidic due to its pollution by sulphur dioxide, etc.

In the early 1950's great problems were caused by sulphur dioxide and the smoke which is always with it. Apart from blackened buildings, and long-term fogs, they caused serious health problems. For example, in 1952 there were believed to be over 4000 deaths caused by air pollution in London alone. Most of these deaths were caused by lung illnesses such as bronchitis.

The Clean Air Act (1956) set up 'smokeless zones' in large towns and cities. In these zones, coal cannot be burned; smokeless fuels having to be used instead. Smokeless fuels have had all the sulphur removed.

Sulphur dioxide is still produced today when coal is burnt in large coal-fired power stations. To reduce sulphur dioxide emissions, the waste gases containing sulphur dioxide are washed with a mixture of limestone and water.
This removes sulphur dioxide and produces gypsum, used for making plasterboard. Although this process is expensive, it reduces the concentrations of sulphur dioxide in the atmosphere.

Probably more significant today is the build up of oxides of nitrogen in the atmosphere. Much of this is caused during combustion of petrol in car engines. Like sulphur dioxide, oxides of nitrogen in the atmosphere produce acid rain. Up to 90 per cent of the oxides of nitrogen in the atmosphere could be removed if all cars were fitted with catalytic converters. The exhaust gases are forced through the converter which contains a ceramic honeycomb structure. The honeycomb is covered with a thin layer of platinum and other precious metals. Here oxides of nitrogen are converted to harmless nitrogen.

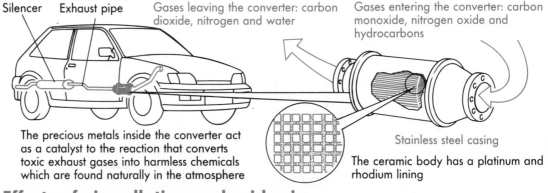

Silencer   Exhaust pipe   Gases leaving the converter: carbon dioxide, nitrogen and water   Gases entering the converter: carbon monoxide, nitrogen oxide and hydrocarbons

The precious metals inside the converter act as a catalyst to the reaction that converts toxic exhaust gases into harmless chemicals which are found naturally in the atmosphere

Stainless steel casing

The ceramic body has a platinum and rhodium lining

Effects of air pollution and acid rain

The effects of air pollution and acid rain include:

1 Damage to stonework on buildings (see page 86).

2 Rivers and lakes over a wide area become more acidic. Increased acidity causes fish to die. Lakes in Sweden and Norway now have no fish. The acidity in the lakes can be neutralized by adding limestone.

3 Forests are seriously damaged. Trees are stunted, needles and leaves drop off and the trees die.

4 Iron and steel structures corrode faster (e.g. iron railings).

ACTION!   **The effects of sulphur dioxide on seedlings**

Look at the shows two experiments set up to show the effect of sulphur dioxide on some cress seeds as they germinate and grow. The seedlings in in A germinate but soon die whilst the seeds in B germinate and grow well.

A dish containing cotton wool and sodium sulphite solution (which releases sulphur dioxide)

The seedlings grow better in **B**

A well-watered pot of cress seedlings

A well-watered pot of cress seedlings

Sulphur dioxide, water and plants

Air

Water and plants

A dish containing cotton wool and water

A   **B**

Polythene bag

42

(a) What can you conclude about the effect of sulphur dioxide on these seedlings?
(b) Experiment B is called a **control**. Explain why a control is necessary.
(c) Why would it not be correct to use a clear plastic bag in A and a black plastic bag in B?

WATER POLLUTION

Water is vital to our lives. It is estimated that we each use approximately 120 litres (26 gallons) of water each day. In addition, many of the things we use require large amounts of water to make them. For example, it takes:

7 litres of water to make 1 pint of beer;
55 litres of water to make a jar of instant coffee;
26 000 litres of water to make 1 tonne of newspaper;
45 000 litres of water to make 1 tonne of steel.

We also produce large amounts of waste water from our baths, washing machines, toilets etc. This water should be cleaned up in sewage works before it is released into the environment. Factories which use large amounts of water are often near rivers. Over 50 000 000 gallons of waste water enter rivers in Great Britain every day from sewage works, factories and farms. Great care has to be taken to ensure that unwanted pollutants are removed from the water.
Serious pollutants include: nitrogen compounds, such as ammonia and nitrates; heavy metals, such as lead and mercury; and phosphates.

ACTION!

Pollution profile

The table below shows some small sea creatures which can exist in water containing different amounts of pollution.

Amt. of pollution	Animals found		
Very high			
High			
Low			
Very low			

A river was examined at four places (W, X, Y, and Z) and the only creatures present in the water are shown in this table.

W	X	Y	Z

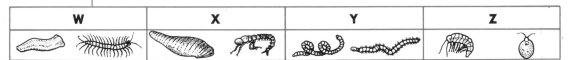

Arrange the four samples in order of pollution with the least polluted first.

Pollution of water by nitrogen compounds

At room temperature about 30 cm^3 of oxygen will dissolve in one litre of water. This small amount of oxygen is all that is available for fish and other river life. Ammonia, added to river water, removes dissolved oxygen from the water. The ammonia is oxidized by bacteria in the river forming nitrates. Since nitrates encourage plant growth, green algae start to grow on the water surface.

These algae prevent light entering the water. When the algae die and decay, more oxygen is removed from the water. Eventually the river becomes completely dead.

Ammonia is present in sewage. Most of the ammonia and nitrates which enter rivers come from the escape of effluent from farms: either as slurry from farm animals; or as water from silage spills. In 1987, 3890 incidents of pollution from farms were reported. Some of the artificial fertilizers, which contain ammonia and nitrates, are washed into rivers.

ACTION!

Fixing nitrogen!

The table shows some of the ways in which nitrogen gets into the soil. (The figures are estimated world figures.)

Route	Nitrogen in the soil (millions of tonnes)
lightning	20
oxides of nitrogen from vehicle exhausts	20
bacterial action in soil	70
bacterial action in water	20
artificial fertilizers	50

Table

(a) Which of these routes are natural and which are not natural?

(b) Express the results in the table as a pie chart.

(c) Calculate the amounts of nitrogen converted both naturally and otherwise.

(d) How is the situation likely to change in the next few years? Explain your answer.

High levels of nitrate in water cannot be removed. These can cause health problems. 'Blue baby syndrome' is a blood disorder in babies of up to six months old caused by the inability of the blood to carry oxygen around the body. It causes the baby to turn blue. It is extremely rare and it is believed that it is linked to high levels of nitrate in water. High levels of nitrate are also believed to cause stomach cancer.

ACTION!

Drinking water

According to a Government report of 1988, a million Britons are drinking tap water containing higher levels of nitrate than permitted by the EC. Recommendations to reduce these levels include: using no artificial fertilizer between mid-September and mid-February; and sowing autumn-sown crops rather than spring-sown ones.

Another report in 1990 said that there were traces of a number of toxic pesticides in tap water in Britain. Bringing drinking water up to the standards required by the EC will be extremely expensive.

Can you suggest any steps which could be taken to improve the quality of our drinking water taken from rivers and lakes?

Pollution by heavy metals

Very serious problems are caused when metals are discharged into rivers. Metals such as lead, mercury and cadmium, kill river life and make the water toxic. Often these poisons are **cumulative**. This means they build up in fish, mammals and so on. High concentrations of mercury can cause blindness, speech defects and inability to co-ordinate movements in animals. It can even cause death. Can you explain how this affects humans?

Pollution by phosphates

Phosphates and detergents get into rivers from washing powders. Phosphates are added to washing powders to help give a whiter wash. Like nitrates, phosphates encourage the growth of plants in water. Detergents are now usually biodegradable to prevent this from occuring.

ACTION!

The effects of detergent on growing plants

Five beakers were set up. Each beaker contained distilled water. No detergent was added to beaker A. To the other beakers, different amounts of detergent were added. Finally, 20 plants were added to each beaker and the beakers were left for one week.

The results are summarized in the table below:

Beaker	Amount of detergent added (cm^3)	Number of browned plants
A	0.0	2
B	0.5	6
C	1.0	9
D	1.5	12
E	2.0	14

Table

(a) What must be kept the same during this experiment?

(b) What can you conclude about the effect of detergent on these plants from this experiment?

Did you know?

Pollution of the River Rhine

In November 1986, there was a serious accident in Switzerland. Following a factory fire, 30 tonnes of agricultural chemicals were washed into the River Rhine. During the next 12 days, they flowed down the River Rhine towards the sea. As they went, they produced a number of problems. These included:

1 killing 34 varieties of fish;

2 the Rivers Waal and Ijssel had to be closed off to prevent dyke and land pollution;

3 alternative sources of drinking water had to be found.

ACTION!

Something which you feel strongly about

Wherever you live there will be something which is affecting the environment near you. Here are some examples:

Open-cast mining There is a plan to extract millions of tonnes of coal from an area of open space in the middle of the city of Stoke-on-Trent.

The area is surrounded by housing. The project will cause considerable mess, noise and discomfort to local residents.

Extending the M3 motorway Plans to extend the M3 motorway in Hampshire have run into opposition because they are cutting through an area of natural beauty at Twyford Down, near Winchester.

Pollution of beaches The quality of sea water at Blackpool, the country's leading resort, may be below acceptable standards for the EC. This is due to untreated sewage being pumped into the sea.

Choose an issue which is important in your area. Talk to friends and your teacher to try to find out as much about it as possible. What can be done to improve your local environment? Can it be done without spending a lot of money? Devise a survey for testing local opinions!

Did you know?

Modern Egypt

Look at the map of Egypt.

For thousands of years only a very narrow strip of land on both sides of the River Nile has been cultivated.

The rest of the country is largely desert.

The fertile strip was flooded every year by the River Nile.

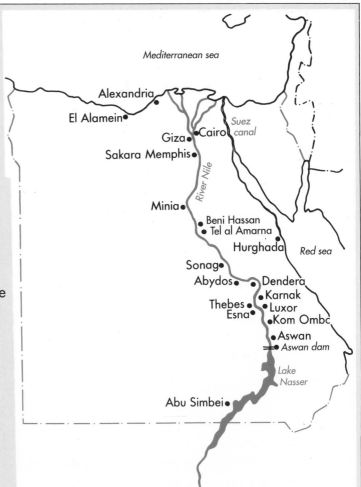

However, in 1980 the new Aswan Dam was built which held back the water and used it to generate electricity. This provides all of the electricity needed by the country for developing industry. However, the Nile no longer floods. The fertile strip has to be irrigated with water from the river and artificial fertilizers have to be used. There is also less water in the river and so any pollution is not as diluted as it otherwise would be.

UNIT 6
TYPES AND USES OF MATERIALS

You should already be familiar with a wide range of different materials and be able to describe simple properties (e.g. stone is hard and cold to touch whereas wood is usually soft and warm). You should be able to group materials according to these simple properties. You should know that on heating and cooling materials may change. Water turns to steam on heating and to ice on cooling. These changes are temporary. When heated in air, wood may burn. This change is permanent.

MATTER AND MATERIALS

In science, a number of words are used with very similar meanings. The words **matter** and **materials** are words which mean the same as substance or stuff. Materials come in all sorts of shapes, sizes and types and they have very different **properties**.

The use we make of various materials depends very much on their properties. Copper, a metal, is used for electrical wiring because it is a good conductor of electricity.

The table below lists 12 different properties and gives examples of extremes.

Property	Examples	Description
hardness	concrete, wool	hard, soft
strength	steel, paper	strong, weak
flexibility	wood, rubber	stiff, flexible, bendy
density	lead polystyrene	dense, feels heavy feels light
colour	coke, aluminium	black, silver
texture	mortar glass	rough smooth
solubility in water	salt sand	dissolves does not dissolve
smell	chlorine air	smells like a swimming pool no smell
conductivity of heat	copper feathers	good conductor of heat poor conductor of heat
conductivity of electricity	copper	good conductor of electricity
transparency	glass wood	transparent (can be seen through) non-transparent
corrosion	gold iron	does not corrode corrodes or rusts

These 12 properties can be used to describe a particular material. For example, a brick could be described as a hard, strong, stiff and dense solid. It is usually reddish-brown and rough in texture. It is insoluble in water and has no smell. It is a poor conductor of heat and electricity. It is opaque and does not corrode. Materials can often be identified from such a description.

ACTION!

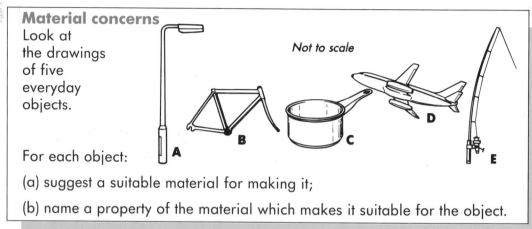

Material concerns

Look at the drawings of five everyday objects.

Not to scale

For each object:

(a) suggest a suitable material for making it;

(b) name a property of the material which makes it suitable for the object.

Material strengths

When choosing suitable materials for a particular purpose, it is often wise to test the strength of possible materials. It is necessary to test samples that are the same size and shape and to test them in the same way.

A fair test of strength will only be made if all these **variables** are kept the same. Here are shows three ways of testing the strengths of different materials:

(a) crushing; (b) bending; (c) stretching.

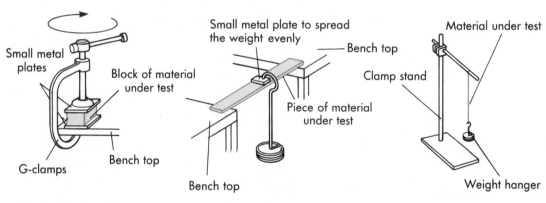

Crushing: Compare different materials by comparing the angle through which the handle on the clamp turns

Bending: Compare the amount of bending in different materials

Stretching: Compare the extensions of different materials for a given weight

Concrete is a very common building material used for bridges, beams, etc. It is not a very strong material, however. A downward force on a concrete beam will cause it to break.

Horizontal steel reinforcing rods transfer the force sideways and make the concrete beam stronger.

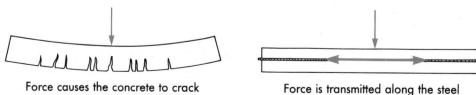

Force causes the concrete to crack

Force is transmitted along the steel reinforcing rod

Stretching different types of 'wool'

At one time, wool was only used for knitting. Nowadays, other materials are used but are still called 'wool'. The graph shows the results of stretching samples of 'wool' labelled A, B and C.

Each piece of 'wool' was 50 cm long and 200 g weights were added up to a maximum of 2000 g.

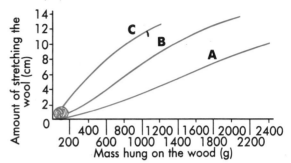

(a) What was the length of A when a 1000 g weight was added?

(b) Which 'wool' stretched most and which least?

(c) Which 'wool' broke before the end of the test?

Material hardness

The Mohs' scale of hardness was developed by Mohs in 1812. It is used to compare the hardness of materials especially rocks and minerals. It was devised by taking diamond (the hardest substance) giving it a value of 10. Other substances were put in order depending upon whether they scratch other substances. A substance with a hardness of 6 will scratch a substance with a hardness of 5 and will be scratched by a substance with a hardness of 7.

Testing the hardness of minerals

A geologist, working in the field, is able to find the approximate hardness of minerals using the following tests. The hardened steel blade of a penknife has hardness of about 6. A 2p coin has a hardness of 3.5 and a fingernail a hardness of 2.

What can you conclude about four minerals, A, B, C, and D from the results in table 1:

Mineral	Mineral scratched by		
	Steel blade	2p coin	Fingernail
A	√	√	X
B	X	X	X
C	√	X	X
D	√	√	√

Table 1

Table 2 lists the Mohs' scale of hardness.

Mohs' scale value	Material	Mohs' scale value	Material
10	diamond	5	apatite
9	corundum	4	fluorite
8	topaz	3	calcite
7	quartz	2	gypsum
6	feldspar	1	talc

Table 2

Using table 2, suggest an identity for each material.

THE STATE OF MATTER

There are different ways of dividing materials into groups. One way is to divide them according to their states. Matter or materials can exist in three different forms or **states**.

Water is a very common substance. You will know that water can exist in three forms; **ice**/solid; **water**/liquid; **steam**/gas.

Sometimes water in a gas form is called water vapour. Can you say what the difference is between steam and water vapour?

When liquid water is heated, it turns to steam at 100°C. The water is said to be **boiling** and this temperature is called the **boiling point** of water.

When steam is cooled down, it turns back to water. You will have seen the water which forms on a cold window in a steamy kitchen. This change back from steam to liquid water is called **condensation**.

When water is cooled, it turns to ice at 0°C. This is called **freezing** and 0°C is called the **freezing point** of water. At 0°C, **melting** of ice also takes place and ice turns to liquid water.

Sometimes steam (or water vapour) can turn directly into a solid.

This happens inside a freezer.

Solid ice forms inside the freezer when the steam in the air rapidly cools.

This change is called **sublimation**.

These changes of state are summarized opposite.

Similar changes take place with other substances.
All substances can exist in three states of matter depending upon conditions. These are solid, liquid and gas.

The typical properties of solids, liquids and gases are compared in table 1.

Property	Solid	Liquid	Gas
volume	definite	definite	fills the whole container
shape	definite	takes up shape of bottom of the container	takes up shape of whole container
density	high	medium	low
expansion on heating	low	medium	high
ease of compression	very low	low	high
movement of particles	very slow	medium	fast moving particles

Predicting the state of a substance

Table 2 compares the melting and boiling points of some common substances. Assume for this exercise that room temperature is 20°C.

Substance	Melting point °C	Boiling point °C
hydrogen	−259	−253
nitrogen	−214	−196
oxygen	−219	−183
ethanol	−117	−78
ammonia	−78	−33
mercury	−39	357
bromine	−7	58
sodium	78	890
iodine	114	183

[continued on page 51]

sulphur	119	445
zinc	419	908
potassium chloride	776	1427
sodium chloride	801	1420
copper	1083	2582
iron	1539	2887

A substance will be a solid at room temperature, if its melting and the boiling point are above 20°C. Look at the list of substances in table 2.

Which of these substances are solid at room temperature?

A substance will be a liquid at room temperature if the melting point is below 20°C but the boiling point is above 20°C.

Which substances in table 2 are liquid?

A substance is a gas at room temperature if both the melting and boiling points are below 20°C.

Which substances in table 2 are gases?

You should now be able to work out whether any substance is a solid, liquid or a gas, at a given temperature provided that you are given its boiling and freezing points.

Elements, mixtures and compounds

Materials can be divided into groups depending upon whether they are **elements, mixtures** or **compounds**. These will be considered in Unit 8.

DISSOLVING

One property listed in the table on page 47 is the ability to dissolve in water.

When salt is added to water and the mixture is stirred, the salt dissolves and forms a salt **solution**. The salt disappears from view. However, it is still there because the water tastes salty.

A substance which dissolves to make a solution is said to be **soluble**.
A substance which does not dissolve is said to be **insoluble** (e.g. sand).
The substance which is dissolved is called the **solute** (e.g. salt, in this case) and the substance which does the dissolving is called the **solvent** (e.g. water).
Water is the commonest solvent but there are many others (e.g. ethanol, hexane).

The following rules about solubility are worth remembering:

1 Most solids dissolve better in hot water than in cold water.

2 Liquids either dissolve in water (**miscible**) or do not mix with water (**immiscible**). Changing the temperature does not affect this.

3 Gases dissolve better in cold water than in hot water.

Solubility

If you keep dissolving a solute in water at a particular temperature you will reach a stage where no more will dissolve. The resulting solution, which has the maximum amount of solute dissolved at a particular temperature, is called a **saturated solution**.

The **solubility** of a solute is the number of grams of the solute which will dissolve in 100 g of water at a particular temperature.

Most solids dissolve better in hot water than cold water.

The graph shows how the solubility of various solids changes with temperature.

These graphs are called **solubility curves**.

If a hot saturated solution is cooled, the solute will crystallize out.

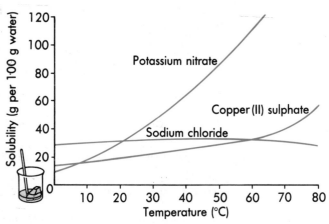

Solubility curves

(a) Which solid is most soluble at: (i) 10°C; (ii) 60°C?

(b) Which solid has almost the same solubility at all temperatures?

(c) What mass of copper (II) sulphate would form a saturated solution in 50 g of water at 20°C?

(d) A solution containing 100 g of potassium nitrate at 60°C is cooled at 20°C. What would you see as this happened?

(e) At which temperature are the solubilities of sodium chloride and potassium nitrate the same?

Solu-ability

The table shows the solubilities of some substances in water at 25°C:

	Substance	Solubility (g per 100g water)
solids	sodium nitrate	92
	potassium chloride	96
	potassium nitrate	36
	ammonium nitrate	214
	ammonium chloride	39
gases	ammonia	52.5
	carbon dioxide	0.17
	hydrogen chloride	72
	oxygen	0.0045
	nitrogen	0.0019

(a) Which solid in the table is least soluble in water at 25°C?

(b) Name the two gases which are very soluble in water.

(c) One of the gases in (b) dissolves in water to form an acid solution and the other forms an alkaline solution. How could you show which was acid and which was alkaline?

ACIDS AND ALKALIS

Many of the substances around us are acids or alkalis. The sharp taste we get when we bite into an apple is an acid. Acids always have a sour taste, although we would be unwise to taste most of them! Acids are present in lemons, oranges and limes (i.e. citrus fruits contain citric acid). The sourness in sweet and sour chicken comes from vinegar. This contains ethanoic (acetic) acid.

There are three common mineral acids: hydrochloric acid: sulphuric acid; nitric acid.

Washing powders, caustic soda and ammonia solution are examples of alkalis. An alkali solution is usually soapy.

A **neutral** substance is neither acid nor alkaline (e.g. pure water, ethanol or petrol).

The simplest way of testing for an acid or an alkali is to use litmus paper or litmus solution. Litmus is a purple-coloured extract of a lichen which changes colour depending upon whether acid or alkali is added. If litmus is added to an acid, the litmus turns red. If it is added to an alkali it turns blue.

This can be summarized as:

Alkali	Acid
Blue	Red

Litmus gives you an idea about the strength of an acid or alkali. Both vinegar and sulphuric acid turn litmus red.

ACTION!

Testing for acids and alkalis

Three test tubes contain different liquids. A piece of red litmus paper and a piece of blue litmus paper are added to each test tube. The results are summarized in the table below.

Liquid	Red litmus	Blue litmus
A	stays red	turns red
B	turns blue	stays blue
C	stays red	stays blue

What can you conclude about each liquid from these tests?

Universal indicator is a better test for acids and alkalis. This indicator is a mixture of other simple indicators and it changes through several colours. From the colour you can work out the pH. The colours are shown in this table:

pH	Colour	Acidity/alkalinity
1		
2		
3	red	
4		acid
5	orange	
6	yellow	
7	green	neutral
8	blue	
9	indigo (blue/violet)	
10		
11		alkali
12	purple	
13		
14		

The pH is a number on a scale which shows how acid or how alkaline a substance is. A substance which is neither acid nor alkaline is said to be neutral. If a solution is slightly acidic, it would turn Universal indicator yellow and have a pH of 6. It is possible to measure the pH of a solution using a pH meter.

Measuring pH

Suppose you wanted to measure the pH of:
(a) blackcurrant cordial; (b) bleach.
Why would these measurements be difficult using universal indicator?

PURE CHEMICALS AND PURIFICATION

A pure chemical is chemical which does not contain **impurities**. A pure chemical has a definite melting point. The presence of impurities lowers the melting point and causes the substance to melt over a range of temperatures.

There are a number of methods which can be used to produce pure chemicals. The method used has to be chosen carefully for each purification.

Filtration and evaporation

An impure form of salt found underground in Cheshire is called rock salt. This consists of salt mixed with impurities, such as sand, which do not dissolve in water and are thus insoluble. The fact that the impurities do not dissolve in water is the basis of a method used to purify rock salt.

The rock salt is crushed using a pestle and mortar. The crushed rock is added to water and the mixture is stirred.

The salt dissolves but the impurities sink to the bottom and form a **sediment** or **residue**.

The salt solution can be removed by **decanting**.

— Insoluble substance

— Salt solution

Alternatively, the salt solution can be removed by **filtering**. In the kitchen, flour may be sieved to remove any lumps. The flour passes through the small holes in the sieve but the lumps do not. Filtering is a very similar process. A filter paper has many very small holes through it.

The filter paper is folded into a cone shape and placed in a funnel to support it.

The mixture of salt solution and solid impurities is poured into the funnel. The solid impurities remain on the filter paper and the solution passes through to be collected in a beaker.

The solution collected in the beaker is called the **filtrate**.

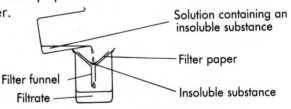

Solution containing an insoluble substance

Filter paper

Filter funnel

Filtrate

Insoluble substance

Solid salt can be recovered from the salt solution by **evaporation**.

The solution is heated in the apparatus shown until all of the water has boiled away.

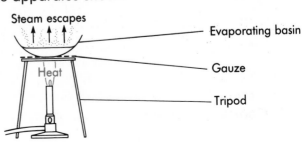

Steam escapes

— Evaporating basin

Heat

— Gauze

— Tripod

Evaporating all of the water away is called **evaporating to dryness**. It does not produce good crystals, however. To get good crystals the solution should be evaporated until a small volume of solution remains. The solution should then be left to cool when crystals will form.

Obtaining a solvent from a solution

Evaporation is used to obtain a solute from a solution (e.g. salt). In some cases, it is important to recover the solvent from a solution. This can be done by a process called **distillation**. This is really evaporation followed by condensation.

Look at the apparatus set up to recover some water from ink. The ink is boiled and the steam is condensed. The liquid produced is called the **distillate**. If this experiment is carried out carefully pure water can be produced. Even so it is difficult to stop the ink boiling over and to condense all of the steam.

The diagram opposite shows an improved apparatus which uses a **condenser**.

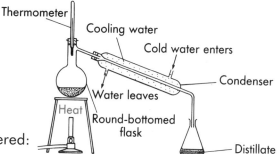

This piece of apparatus condenses the steam efficiently.

Cold water passes through the condenser to cool down the steam.

The following points should be remembered:

1 Only steam leaves the condenser. The other substances in the ink remain in the flask.

2 The thermometer measures the temperature of the steam. The bulb of the thermometer is level with the side-arm of the flask and not dipping in the ink. The maximum temperature recorded on the thermometer when ink is distilled should be 100°C.

3 The condenser consists of two tubes (one inside the other). Steam passes through the inner tube and cooling water passes through the outer tube. The cooling water enters at the bottom of the condenser and leaves at the top.

4 The condenser must slope downward so that the water which condenses runs into the receiver.

5 The receiver should be open at the top, (i.e. there should be no cork in it).

Separating mixtures of liquids

Ethanol and water mix together completely to form a single solution and are called miscible.

Hexane and water do not mix well. They form two separate layers. The top liquid is almost completely hexane whilst the lower liquid is almost completely water. They are said to be immiscible. The liquid in the lower layer has a greater density than the liquid in the upper layer.

Immiscible liquids can be separated using a separating funnel. A mixture of hexane and water is placed in the funnel.

After standing the tap is opened.

The water layer runs out through the tap.

The hexane layer remains in the funnel and can be run out into another beaker.

Miscible liquids are much more difficult to separate but their separation is important in industry.

Mixture of miscible liquids can be separated by **fractional distillation** if the boiling points of the liquids are not too close together.

A mixture of ethanol (boiling point 78°C) and water (boiling point 100°C) can be separated by fractional distillation.

The diagram shows apparatus suitable for the fractional distillation of a mixture of ethanol and water in the laboratory.

The mixture to be separated is placed in the flask containing small pieces of broken china to ensure that the liquid does not boil over when heated.

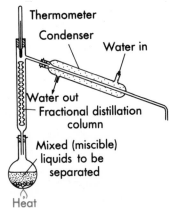

The flask is heated slowly with receiver number 1 in place.
The ethanol starts to boil first because it has a lower boiling point.
The vapour passes up the fractional distillation column. Any water vapour which gets into the column at this stage condenses and drops back into the flask.
The temperature on the thermometer remains below 80°C and only ethanol distils over. The liquid collected in the first receiver is called the first **fraction** and consists almost entirely of ethanol.

When the temperature reaches 80°C, receiver 2 is put in place and the temperature rises quickly to 95°C. A second fraction is collected.

When the temperature reaches 95°C, receiver 3 is put in place and soon a large volume of liquid collects in the receiver. The results of this experiment are shown in the diagrams below.

ACTION!

Naming fractions

What do you think fractions 1, 2 and 3 consists of?

Fractional distillation is used in industry in order to:

1 separate oxygen and nitrogen from air by fractional distillation of liquid air, (oxygen and nitrogen are important chemicals used in the manufacture of steel and ammonia, respectively).

2 refine crude oil to produce valuable products that can be used as fuels or starting chemicals for making plastics.

3 concentrate ethanol in whisky production.

Chromatography

Chromatography is relatively simple. It is used to separate mixtures of substances dissolved in a solvent. It can also be used to identify substances. The simplest form of chromatography is called **paper chromatography**.

Paper chromatography is often used to separate mixtures of inks and dyes. It relies upon the different rates at which the dyes spread across a piece of filter paper.

A spot of the mixture of dyes is placed in the centre of a piece of filter paper. When the solvent travels up the wick it reaches the spot. The spot spreads out on the piece of filter paper. Each dye spreads out at a different rate depending upon the relative liking of the dye for the solvent and the paper. Each dye in the original mixture produces a different ring. In this example the original mixture contains two dyes.

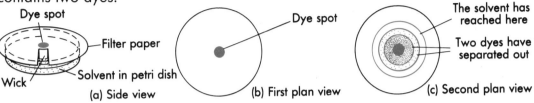

In practice, a square sheet of filter paper is often used. The sample spots are put on the base line and the paper is dried thoroughly. The paper is then coiled into a cylinder and put into a tank with a lid.

At the bottom of the tank is a small amount of solvent. The solvent travels up the filter paper and the spots are separated. When the solvent has nearly reached the top of the filter paper, the paper is removed and the position of the solvent marked.

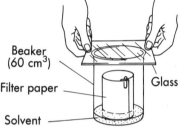

The paper is dried.

Look at this example of this form of chromatography. Explain how the substances are related.

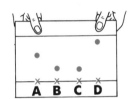

× shows the original position of each substance
● shows the final position of each substance

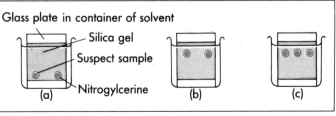

Chromatography can be used to identify the dyes in food to ensure they are permitted dyes. It can also be used to diagnose medical problems. For example, some people are unable to use the amino acids that they eat to build up proteins (*see* page 36). By carrying out chromatography on their urine, the excess of amino acids leaving the body can be detected.

THE REACTIVITY SERIES OF METALS

One of the properties considered in the table on page 47 was reactivity. Metals can be listed in order of decreasing reactivity and is called the **reactivity series**. The reactivity series for common metals is:

potassium

sodium

calcium

magnesium

aluminium

zinc

iron

lead

copper

silver

gold

Not to scale

The reactivity series is useful for predicting likely chemical reactions.
For example, if iron filings are added to blue copper (II) sulphate solution, a brown solid is formed and the solution turns colourless. The iron and copper change places and free copper is formed.

copper (II) sulphate + iron → copper + iron (II) sulphate

This is called a **displacement reaction**. It takes place because iron is more reactive than copper. Iron is higher in the reactivity series than copper.

If silver powder is added to copper (II) sulphate solution there is no reaction because silver is less reactive than copper.

The Thermit reaction is a practical application of a displacement reaction.
It is used to weld together long lengths of railway track from shorter lengths.
A mixture of aluminium powder and iron (III) oxide powder is placed between the two ends of railway track. The mixture is set alight and a reaction takes place.

iron (III) oxide + aluminium → iron + aluminium oxide

The heat generated is enough to melt the iron which flows into the gap between the rails and welds them together.

ACTION!

Getting a reaction?

For each of the following pairs of chemicals, state whether or not a displacement reaction could take place. In some cases, heat may be needed. If a reaction takes place, name both of the products:

(a) zinc and copper (II) sulphate solution

(b) copper and silver nitrate solution

(c) copper and aluminium sulphate solution

(d) magnesium and copper (II) oxide

(e) iron and zinc oxide

Hypothesis and theory

A **hypothesis** is a suggested explanation for some observations or facts. It is not necessarily exactly right but it does give a basis for further investigation. After testing, the hypothesis may be altered to meet new findings or it may be accepted completely. Once a hypothesis is proved, it becomes a **theory**.

ACTION!

The acid test

Jane was carrying out some experiments with magnesium, zinc and manganese (three metals) and dilute sulphuric acid.

Here are her results.

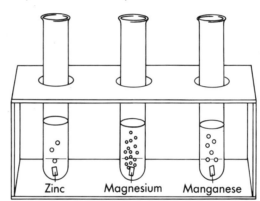

Zinc   Magnesium   Manganese

On the basis of these results she concluded that:

all metals react with dilute sulphuric acid to produce bubbles of colourless gas.

This is her hypothesis. What should she do to test her hypothesis to see if it is true in all cases? If tests prove her hypothesis, it could become a theory.

59

UNIT 7
MAKING NEW MATERIALS

This Unit does not assume that you have any knowledge or experience from key stages 1 and 2.

CHEMICAL REACTIONS

When new substances (called **products**) are made from reacting substances (called **reactants**) a **chemical reaction** takes place.

A chemical reaction is often expressed by a word equation. In the equation the reactants are on the left hand side and the products are on the right hand side. The arrow shows the direction of the reaction. For example:

magnesium + hydrochloric → magnesium + hydrogen
acid   chloride

Although the equation gives the reactants and products it gives no indication of the speed of the reaction.

Oxidation and reduction reactions

Oxidation takes place when oxygen is added or hydrogen is removed during a chemical reaction. Reduction is the opposite of oxidation (i.e. oxygen is removed or hydrogen is added). For example:

1 oxidation magnesium is oxidized (oxygen is added).

magnesium + oxygen → magnesium oxide

2 reduction ethene is reduced (hydrogen is added).

ethene + hydrogen → ethane

Very often, oxidation and reduction take place together. A reaction in which oxidation and reduction take place is called a **redox** reaction. For example, lead (II) oxide heated in a stream of dry hydrogen:

lead (II) oxide + hydrogen → lead + water

Lead (II) oxide is reduced to lead since oxygen is lost. Hydrogen is oxidized as oxygen is added. Hydrogen is called the **reducing agent** because it brings about the reduction of lead (II) oxide. Lead (II) oxide is called the **oxidizing agent** since it brings about the oxidation of hydrogen.

Hydrogen enters

Water goes out

Lead oxide powder turns into lead

Combustion reactions

A **combustion reaction** is a reaction where a substance combines with oxygen and produces energy. When a combustion occurs **oxides** are formed.

Combustion reactions are oxidation reactions. For example, burning carbon in oxygen or burning magnesium in oxygen:

carbon + oxygen → carbon dioxide

magnesium + oxygen → magnesium oxide

When hydrocarbons burn in oxygen, different products are possible. If the hydrocarbon is burnt in excess air or oxygen, carbon dioxide and water are produced.

However, if the hydrocarbon is burnt in a limited amount of air or oxygen, carbon monoxide and water are produced.

Carbon monoxide is very poisonous because it combines with haemoglobin in the blood forming carboxy-haemoglobin. This prevents oxygen being transported around the body and leads to death. Good ventilation is necessary in a room with a gas fire to avoid carbon monoxide formation.

Can you explain why with the help of the drawings?

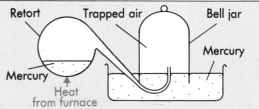

Retort   Trapped air   Bell jar

Mercury

Mercury

Heat from furnace

In the 18th century, the accepted theory of combustion was called the Phlogiston Theory. According to this theory, when combustion took place the substance burning lost a substance called phlogiston. Lavoisier proved that this theory was not correct. When a substance burns it increases in mass if all the products are collected. The substance gains oxygen when it burns. The very careful work of Lavoisier disproved the Phlogiston Theory and established the new theory of combustion which we use today.

Lavoisier was born in Paris in 1743 and was given an excellent education by his wealthy parents. He studied mathematics, astronomy, chemistry and botany at Nazarin College.

He graduated in law but was appointed Director of the Academy of Sciences in 1768. His scientific work continued here. Some of his experiments led to an improved way of making gunpowder.

He became a politician and was elected to the Assembly of Orleans in 1787. He strived to improve the living conditions for ordinary people.

However, he became a target because of his wealth when the French Revolution started. In 1794 he was arrested, tried by a revolutionary court in just a few hours, and sentenced to death along with 27 others.

On the same evening, the man who did so much to improve the lives of ordinary people and advance scientific thought was guillotined in front of the masses and buried in a common grave.

Lavoisier's studies into combustion required very accurate measurements of masses and skill in handling gases and collecting them over mercury or water.

(a) The photograph on page 62 shows the type of balance used by Lavoisier for weighing. What problems would this kind of balance produce?

(b) When mercury was heated in contact with air in the apparatus above, oxygen was taken in by the mercury to form mercury oxide.

(i) Describe what happened
   to the mercury level in the bell jar.

(ii) The mercury oxide produced was
   removed from the apparatus.
   When the mercury oxide was produced,
   it decomposed and formed oxygen gas.
   Draw a diagram of apparatus which
   could be used to collect the
   oxygen produced.

Respiration

Respiration is very similar to combustion. The fuel this time is food and this is oxidized in the cells of the body to produce carbon dioxide and water. Energy is also produced and this enables us to do work and keeps up our body temperature.

If the food used is glucose, the word equation is:

glucose + oxygen → carbon dioxide + water + energy

The carbon dioxide and water vapour are transported back to the lungs and breathed out.

The air we breathe out contains more carbon dioxide and water vapour than the air we breathe in.

Photosynthesis

Photosynthesis is the opposite of respiration. Respiration and combustion both use up oxygen. Fortunately, the oxygen used up is replaced by **photosynthesis** in green plants. Green plants take in carbon dioxide through their leaves. In sunlight, plants produce carbohydrate. The other product is oxygen which escapes into the atmosphere:

energy + carbon dioxide + water → carbohydrate + oxygen

Green plants are essential for keeping the oxygen level in our atmosphere constant. One fifth of the oxygen is produced in the rain forests of South America which are being steadily reduced in area. Increased levels of carbon dioxide in the atmosphere lead to the earth warming up very slowly and this is called the **greenhouse effect**. Can you think of more than one reason for conserving the rain forests?

Decomposition reactions

A substance is said to **decompose** when it splits up into two or more new substances.

Some substances decompose slowly on standing. For example, hydrogen peroxide decomposes slowly at room temperature:

hydrogen peroxide → water + oxygen

When the decomposition occurs on heating, it is called thermal decomposition. Examples of thermal decomposition include:

1 heating calcium carbonate

calcium carbonate → calcium oxide + carbon dioxide

2 heating copper (II) sulphate crystals

hydrated copper (II) sulphate → anhydrous copper (II) sulphate + water

Substances which cannot be decomposed by heating may be decomposed with electricity. This is called **electrolysis**. The substance being decomposed must be molten or in solution. For example, molten lead (II) bromide can be decomposed into lead and bromine by electrolysis using the apparatus shown:

lead (II) bromine → lead + bromine

Lead is produced at the negative electrode (cathode) and the bromine is produced at the positive electrode (anode).

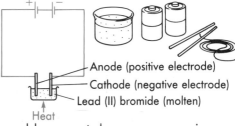

Anode (positive electrode)
Cathode (negative electrode)
Lead (II) bromide (molten)
Heat

Neutralization reactions

All acids contain hydrogen. This can be replaced by a metal or an ammonium ion. The substance formed when the hydrogen in an acid is replaced is called a **salt**.

When an alkali is added to an acid, the acidity is slowly destroyed. If equal amounts of acid and alkali are mixed together a neutral solution is formed.

This process is called **neutralization**. For example, sodium chloride (commonly known as salt) is produced when sodium hydroxide (alkali) neutralizes hydrochloric acid:

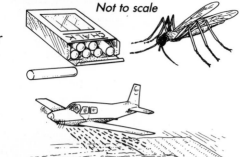
Not to scale

$$\text{sodium hydroxide} + \text{hydrochloric acid} \rightarrow \text{sodium chloride} + \text{water}$$

There are many everyday examples of neutralization:

1 All humans have several hundred cubic centimetres of hydrochloric acid in the gastric juices of the stomach. This is used in the digestion of food. Minor problems of indigestion are caused by excess acid in the stomach. This can be corrected by taking antacids such as milk of magnesia, (a suspension of magnesium hydroxide) or bicarbonate of soda (sodium hydrogencarbonate).

2 Lime mortar consists of a mixture of calcium hydroxide and water. This hardens when it absorbs carbon dioxide from the air. The carbon hydroxide is neutralized by acid gases in the air. Calcium carbonate is formed:

$$\text{calcium hydroxide} + \text{carbon dioxide} \rightarrow \text{calcium carbonate} + \text{water}$$

3 Insect bites and stings involve the injection of a small amount of acid or alkali into the skin. This causes irritation. Nettle stings, ant bites and bee stings involve the injection of acid. The sting or bite should be treated with calamine lotion (a suspension of zinc carbonate) or bicarbonate of soda, to neutralize the acidity and remove the irritation.

4 Many inland lakes in Scotland and Scandanavia are slowly becoming acidic because of air pollution and acid rain. Fish are dying and lakes are becoming lifeless. In an attempt to correct this, the land around the lakes is being treated with lime. As the lime is washed into the lakes it neutralizes some of the acidity.

5 Farmers have to control the pH of their soil. If the soil becomes too acidic, a good yield of crops cannot be obtained. Rain and artificial fertilizers tend to make the soil more acidic. The farmer can neutralize land by treating it with lime.

Can you think of any further examples?

The corrosion of metals

One of the disadvantages of some metals is their tendency to corrode; an expensive problem.

When a metal corrodes, it reacts with oxygen and water in the air.

Corrosion is an oxidation process:

$$\text{metal} + \text{oxygen} \rightarrow \text{metal oxide}$$

Generally, there is a relationship between the position of a metal in the reactivity series and the way it corrodes. The higher a metal is in the reactivity series the more reactive it is, and the more quickly it will corrode.

Very reactive metals such as potassium and sodium corrode very quickly. They are stored under paraffin oil to prevent them coming into contact with air.

Corrosion of iron and steel is usually called **rusting**. It costs hundreds of

Anhydrous calcium chloride

2

Oil

4

millions of pounds each year in Great Britain. The diagram shows an experiment to find out what causes rusting of iron and steel to take place.

Test tube 1 An iron nail is put into water. The nail is in contact with air and water. Rusting takes place.

Test tube 2 Anhydrous calcium chloride removes all the water vapour in the air. The nail is in contact with air but not water. Rusting does not place.

Test tube 3 The distilled water is boiled before use to remove any dissolved air. The nail is in contact with water but not air. Rusting does not take place.

Test tube 4 The nail is in oil. It is not in contact with air or water. No rusting takes place.

From these experiments, it can be concluded that air and water have to be present before rusting of iron and steel takes place. It can be shown that it is the oxygen in the air which is necessary for rusting. Other substances such as carbon dioxide, sulphur dioxide and salt speed up rusting. Can you find out why?

ACTION!

> ### Rusting of iron uses up oxygen
>
> You are given the following: a burette; trough; iron wool and water.
>
> Describe how you would prove to a friend that it is the oxygen in the air which is used up when iron rusts.

Rusting of iron and steel can be reduced by:

1 **Oiling or greasing** e.g. keeping the lawnmower blades oiled over the winter

2 **Painting** e.g. iron railings

3 **Coating with plastic** e.g. washing-up racks

4 **Coating with zinc: galvanizing** e.g. metal dustbins

5 **Sacrificial protection** If a reactive metal, such as magnesium, is kept in contact with the iron, the magnesium corrodes instead of the iron.

Although magnesium is expensive, this is a good method for preventing rust on the hull of a ship.

ACTION!

> ### Going a bit rusty
>
> The table shows how certain substances affect the rusting of steel.
> A tick indicates that the substance is present and a cross indicates the substance is absent.
>
> | Speed of rusting | Substances present | | | |
> | --- | --- | --- | --- | --- |
> | | water | air | salt | mud |
> | nil | √ | X | X | X |
> | nil | X | √ | X | X |
> | slow | √ | √ | X | X |
> | fast | √ | √ | √ | X |
> | very fast | √ | √ | √ | √ |

Using the table:

(a) Which two substances have to be present for rusting to take place?

(b) Which substances together produce very fast rusting?

(c) Explain why rusting takes place quicker:
    (i) when the paint surface has been scratched?
    (ii) in summer rather than winter?
    (iii) in the exhaust system rather than the rest of the car?

Fermentation

Fermentation is an important industrial process which converts sugar or starch solution into ethanol and carbon dioxide gas. This takes place when enzymes in yeast act on the solution at room temperature. Fermentation continues until about 10 per cent ethanol is present. The enzymes are then poisoned by the ethanol.

Fermentation is used to prepare wine from grape juice, or beer from hops.

Fractional distillation of a solution of ethanol (in water) produces a **spirit**.

ACTION! | Trouble brewing

John and Jenny have hit upon a good idea for a party. They will make ginger beer by fermenting a sugar solution with yeast. Ginger is added to flavour the product.

They left the mixture to ferment in the airing cupboard, adding ginger and sugar from time to time. After two weeks they bottled the ginger beer.

They were waiting to give it to their friends at the party.

However, to their horror the bottles exploded. Can you explain why this might have happened?

ENERGY CHANGES IN CHEMICAL REACTIONS

In a chemical reaction there is often an energy change. In a combustion reaction, or in respiration, energy is produced. A reaction which produces energy is said to be **exothermic**. For example:

sodium hydroxide + hydrochloric acid → sodium chloride + water

In photosynthesis, energy is taken in from sunlight. A reaction which takes in energy is called an **endothermic** reaction. For example, the mixing together solutions of sodium carbonate and calcium nitrate:

calcium nitrate + sodium carbonate → calcium carbonate + sodium nitrate

FACTORS AFFECTING THE RATE OF A REACTION

A reaction which takes place quickly is called a **fast** reaction and is finished in a short time. There are a number of ways of speeding up a chemical reaction:

1 Increasing the surface area of the solid Small lumps of a chemical have a much larger surface area than a single lump of the same chemical of equal mass. Powders have a very large surface area.

Flour dust in a flour mill has to be carefully controlled because mixtures of flour dust and air can explode.

2 Increasing the concentration of reacting substances Doubling the concentration of one of the reacting substances will often double the rate of reaction (i.e. halve the time taken for the reaction).

In reactions involving gases, the concentration can be increased by increasing the pressure.

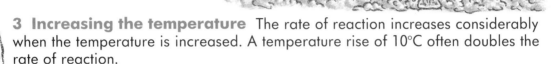

3 Increasing the temperature The rate of reaction increases considerably when the temperature is increased. A temperature rise of 10°C often doubles the rate of reaction.

The rate of souring of milk or spoiling of food is reduced by cooling. A refrigerator, or deep freezer, cools the food down so that the chemical reactions which lead to spoiling are slowed down.

4 Using a catalyst A catalyst is a substance which alters the rate of a chemical reaction without being used up. Usually a catalyst is used to speed up reactions. For example, in the contact process to produce sulphuric acid, the catalyst is vanadium (V) oxide:

sulphur dioxide + oxygen → sulphur trioxide

Sometimes a catalyst is used to slow down a reaction.

For example, additives are added to food to prevent it going bad.

Many chemical reactions taking place in living things are controlled by biological catalysts called **enzymes**. Enzymes are proteins. They have specific properties:

1 A particular enzyme will only catalyze particular reactions, not all reactions.

2 They only work over a limited range of temperatures (e.g. enzymes which operate in the human body will work at temperatures around the normal body temperature of 37°C).

Examples of enzymes are: amylase in saliva which breaks down large starch molecules into smaller glucose molecules; enzymes in biological washing powders which remove stains in cool water; and enzymes in yeast which convert sugar into ethanol during fermentation.

BEFORE   AFTER

THE EARTH'S RESOURCES

The raw materials used in industry come from the Earth. Air, water, rocks, living things and fossil fuels, are all raw materials and are often called resources.

Many of the materials we use at home have been made or **manufactured** using resources from the Earth.

Look at some of the materials used in building a house.

Stainless steel — Concrete
Slate — Clay
PVC
Baked clay — Aluminium
Wood — Brass
Steel — Chipboard
Copper — Glass
— UPVC

ACTION!

Resources in reserve

The bar chart shows the reserves of metal ores which are believed to exist in the Earth.

Metals have been divided, for this investigation, into **ferrous** metals and **non-ferrous** metals.

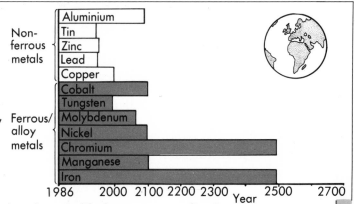

Non-ferrous metals: Aluminium, Tin, Zinc, Lead, Copper

Ferrous/alloy metals: Cobalt, Tungsten, Molybdenum, Nickel, Chromium, Manganese, Iron

1986 2000 2100 2200 2300 Year 2500 2700

(a) Which ferrous metal in the chart is likely to run out first?

(b) Which two ferrous metals will last longest?

(c) Which non-ferrous metals will last longest?

(d) What can be done to make supplies of a metal last longer?

(e) Titanium is used in alloys and when a metal has to withstand high temperatures is needed. Suggest types of materials which might be used in place of titanium.

POLYMERS

Many of the items which used to be made of metals are now made of plastic materials called **polymers**.

Polymers are usually manufactured from petroleum.

For example, a modern car contains many components from polymers.

The parts listed opposite are parts of the chassis and engine made from polymers. Can you list twelve other parts of the car made of polymers.

Polymers are made up from very long chain molecules. These long chains are made up by joining together many small molecules called **monomers**. There can be between 1000 and 50 000 monomer molecules linked together in a polymer chain. The diagram below summarizes the process taking place during polymerization.

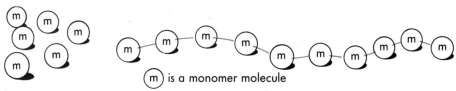

(m) is a monomer molecule

Components commonly made of plastics

An average family car consists of 2730 parts. 771 are made of plastic.

CHASSIS
brake hose
shock absorber fluid
bellows
gaiters

ENGINE
belt drive
fan
timing chain drive and cover
fuel pipes
fuel pump diaphragm
radiator hose
clutch lining

air filter
air filter housing
gaskets
switch gear
cable ducting
carburettor float
cylinder head gasket
fuel tank

PROJECT

1 Fibres for clothing Clothes are made of different fibres. These are usually polymers. They are: natural fibres (e.g. wool, linen, cotton, silk); semi-synthetic fibres (e.g. rayon); synthetic fibres (e.g. nylon, polyester, acrylic).

Look at labels in clothes and complete a table like the one below.

Garment	Fibre	Washing/cleaning instructions
shirt	polyester/cotton	machine wash (ironing not necessary)

Find out from reference books how these different fibres are produced and list as many advantages and disadvantages as you can.

2 Rusting of car bodies Recently, car manufacturers have worked hard to reduce rusting of car bodies. They are so confident that their methods work that they give 'corrosion warranties'.

Collect new car brochures from different manufacturers and find out the methods used to prevent rusting. Find out if there are any conditions associated with corrosion warranties.

Look at cars in a car park and try and identify where corrosion takes place most commonly. From the registration letter you can judge the age of the car.

Try and get hold of a copy of *Which?* magazine that compares the rusting of different makes of car.

UNIT 8
HOW MATERIALS BEHAVE

This Unit does not assume any knowledge or experience from key stages 1 and 2.

MATTER IS MADE UP OF PARTICLES

If you stood in the middle of the desert, you would see nothing but sand stretching in all directions. It would only be when you looked closely, and picked up the sand, that you would realize that sand is not a solid mass but is made up of billions and billions of very tiny grains. Together, the grains of sand look like a solid mass. In a similar way, all matter is made up from very tiny particles called **atoms**.

In 1808, John Dalton proposed that matter was made up of tiny, indivisible particles called atoms, and that atoms of different materials were different to each other.

For example, atoms in a block of iron were different from atoms in a block of carbon.

Carbon atoms   Iron atoms

Atoms could be joined together in different ways to produce all the materials that are known. Later in the 19th century and beyond scientists realized that atoms, themselves, were all made up from even smaller particles called protons, electrons and neutrons.

Electron

Nucleus of protons and neutrons

Did you know?

John Dalton (1766-1844)

In about 420 BC, Democritus, the Ancient Greek philosopher, tried to explain why substances differed in density. Why was a piece of lead very much heavier than a piece of wood of the same size? He thought that less dense substances had more 'open spaces' inside them. This led him to believe that matter was not continuous but was made up of 'pieces'. Borrowing on the ideas of others, he described matter as being made up of 'only atoms and empty spaces'. Unfortunately, Aristotle (384-322 BC), the most important of all Greek thinkers, rejected this idea out of hand and so the theory was discarded. He supported another theory that all things were made from four elements: earth; air; water; and fire, combined together in different proportions. This idea lasted throughout the Middle Ages and survived up until 200 years ago.

The idea of matter being made up of particles was not revived until Dalton's time.

John Dalton was the son of a poor Quaker weaver who lived in Eaglesfield, near Cockermouth in Cumberland. He loved the countryside and liked to study the weather. He was very bright and went to the local Quaker school. He took over there as teacher when he was only 12 years old.

When Dalton was 27, he decided that teaching was getting in the way of his scientific studies. He resigned his teaching post in Manchester and joined the Manchester Literary and Philosophical Society where he devoted much of the rest of his life.

From childhood, Dalton made a daily note of the weather and collected over 200 000 observations on it. Many of Dalton's experiments were rather inaccurate and thus it is his theoretical work that will be remembered, especially his work on the atomic theory. Dalton also studied colour blindness, probably since it affected him personally.

He lived a simple life with regular patterns of work and play. He went on holiday to the Lake District once a year and played bowls every Thursday!

Manchester loved Dalton. At his funeral, 40 000 people filed past his coffin and the public paid for a statue which can still be seen in Manchester Town Hall. There are many other reminders in Manchester of the life of John Dalton.

(a) Scientists often study subjects because they particularly interest them. Describe three things studied by Dalton because they interested him.

(b) Dalton is remembered particularly as a **theoretical** scientist. why do you think this is so?

(c) What would the effects on science have been if Democritus' ideas about particles had not been rejected?

(d) What lesson can be learned, about scientific ideas and theories, from the story of Democritus and Aristotle?

ELEMENTS, MIXTURES AND COMPOUNDS

In Unit 6 we saw how matter could be divided into three groups: solids; liquids; and gases. We can also classify materials as elements, mixtures and compounds.

Elements

All pure substances are made up from one or more of 105 elements.
These are joined together in different ways to give all of the substances in the world around us.

Hydrogen and oxygen are two elements. When hydrogen and oxygen are combined together water is formed.

An element is a pure substance which cannot be split up into anything simpler by chemical reactions. Many of these elements are found in nature but some are made in factories.

The table on page 72 gives some of the common elements. For each element there is a chemical symbol which is one or two letters. Symbols are used as an abbreviation for the element.

Most of the known elements are solids and metals. There are only two liquid elements at room temperature and atmospheric pressure: bromine is a liquid non-metal; and mercury is a liquid metal. The elements in the left-hand column of the table are all metals. Those in the right-hand column are non-metals.

Metallic element	Symbol	Non-metallic element	Symbol
aluminium	Al	bromine	Br
calcium	Ca	carbon	C
copper	Cu	chlorine	Cl
iron	Fe	fluorine	F
lead	Pb	helium	He
lithium	Li	hydrogen	H
magnesium	Mg	iodine	I
potassium	K	nitrogen	N
silver	Ag	oxygen	O
sodium	Na	phosphorus	P
zinc	Zn	sulphur	S

Table

The only elements that are gases at room temperature and atmospheric pressure are: hydrogen; helium; nitrogen; oxygen; fluorine; neon; chlorine; argon; krypton; xenon; and radon.

All elements are made up from tiny particles called atoms. These atoms are so small that they cannot be seen with a microscope.

Mixtures

Elements can be mixed together to form a mixture.

For example, iron and copper powders can be mixed together to form a mixture.

The mixture can be separated with a magnet.

If you look carefully at the mixture with a hand lens you will be able to see pieces of iron and copper. The mixture has all of the properties of iron and copper. Can you think of any other simple ways of separating the mixture?

Compounds

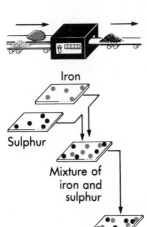

Iron

Sulphur

Mixture of iron and sulphur

Compound of iron and sulphur: iron sulphide

Certain mixtures of elements react together or combine to form compounds. For example, a mixture of hydrogen and oxygen explodes and forms droplets of water.

The formation of a compound from its constituent elements is sometimes called **synthesis**. For example, iron (II) sulphide, the compound formed when the elements iron and sulphur combine, has entirely different properties from iron and sulphur. It is extremely difficult to get iron and sulphur back from iron (II) sulphide. The iron and sulphur atoms join together to form iron (II) sulphide. The synthesis of iron (II) sulphide is summarized in the diagram.

The chemical name of a compound will usually tell you the elements that are combined in the compound. If a compound ends in '-ide', the compound contains only two elements. For example:

sodium chloride   sodium and chlorine
copper (II) oxide   copper and oxygen

There is one important exception to this rule. Sodium hydroxide is composed of **three** elements: sodium; oxygen; and hydrogen.

If a compound ends in '-ate', the compound contains oxygen. For example:

calcium carbonate   calcium, carbon and oxygen
copper (II) sulphate   copper, sulphur and oxygen
sodium hydrogensulphate   sodium, hydrogen, sulphur and oxygen

In a compound, there may be groups of a few atoms (**molecules**) or large arrangements (**giant structures**) of atoms or ions. The following tests help to show which is present:

1 Heating A compound composed of small molecules will have low melting and boiling points. A giant structure of atoms or ions will have a high melting and boiling point.

2 Conductivity of electricity On melting a giant structure of atoms will not conduct electricity but a giant structure of ions will conduct electricity.

Can you discover other ways of testing for these structures?

Methane molecule

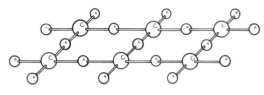

Sand giant structure (silicon dioxide)

ACTION!

Structural problems

The table gives the properties of four substances. From these properties, decide whether the substance is made up from: molecules; a giant structure of atoms; or a giant structure of ions.

Substance	Melting point (°C)	Boiling point (°C)	Electrical conductivity when molten
iodine	114	183	no
sodium chloride	808	1465	yes
silicon dioxide	1610	2230	no
water	0	100	no

Facing the elements, mixtures and compounds

Which one of these diagrams represents:

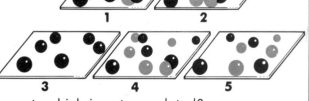

(a) an element?

(b) a pure compound?

(c) a mixture of elements?

(d) a mixture of compounds?

(e) a reaction between two elements which is not completed?

PARTICULAR ARRANGEMENTS

The diagram below shows a simple representation of particles in a solid, a liquid and a gas.

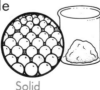

Solid   Liquid   Gas

These drawings are only in two dimensions but they do show some important points:

1 Particles are usually regularly arranged in solids but irregular in liquids and gases.

2 Generally, particles are more closely packed in solids than in liquids and more closely packed in liquids than in gases.

The diagrams, however, cannot show that the particles are moving. In a solid, the particles are not moving very much: they are really only vibrating about fixed points. It is rather like being in a very crowded room and trying to get to the door! In gases the particles are moving rapidly and in all directions. The particles in a gas collide frequently with each other and with the walls of the container. In liquids there is more movement of particles than in solids but less movement than in gases. There is no pattern to the movement of particles in solids, liquids and gases. It is said to be random movement.

ACTION!

Changes of state

Some solid crystals were heated in a test tube, using a water bath, until they melted. A thermometer was put into the liquid and the test tube removed from the water bath.
The test tube and contents cooled. The temperature was recorded every half minute and the results are shown.

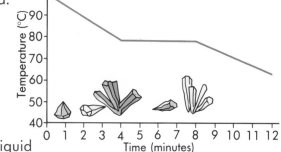

(a) Draw a diagram of the apparatus used:
  (i) to heat the test tube to melt the crystals
  (ii) during the cooling of the liquid

(b) What was the temperature after two minutes?
(c) At which temperature did the liquid turn to a solid?

EVAPORATION

If you leave a saucer of water in a warm room for a couple of days you will find that the saucer is empty. The water has not boiled: evaporation has taken place.

How could you prove this?

Evaporation, like boiling, involves a change from liquid to gas but occurs at any temperature, not necessarily at the boiling point. To understand this, imagine that the saucer of water is made up of millions of tiny water particles that escape into the room as evaporation occurs.

When a liquid boils, the particles are given more energy and they break away from the liquid and move faster. They move apart and occupy more space than in the liquid.

Did you know?

One gram of water (1 cm^3) produces over 1000 cm^3 of steam.

PUTTING A GAS UNDER PRESSURE

If you trap some air in a bicycle pump, it is easy to push in the plunger and compress the gas.

Before    After

The particles are forced closer together.
Gases decrease in volume and increase in pressure
when they are compressed. Do you notice any
other effects when you pump up a bicycle tyre?

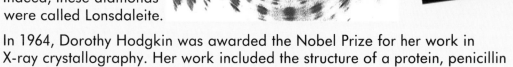

Did you know?

X-ray crystallography

X-ray crystallography is a branch of science
which requires great patience and accuracy.

A beam of X-rays is passed through a crystal. The X-rays are
bent and form a pattern on a photographic plate. From this photograph,
scientists can work out the arrangements of particles in the crystal.

Three of the most famous X-ray crystallographers have been women.
Rosalind Franklin (1921-1958) pioneered X-ray techniques of three-
dimensional crystals. In 1951, she was invited to set up an X-ray unit at the
biophysics department at King's College, London. One of the projects of the
department was to analyse the structure of DNA, a vital chemical which
passes on genetic information from one generation to another. At the same
time, Francis Crick and James Watson were studying DNA. They were
building up models of DNA molecules.

Unknown to Rosalind, one of her papers and her best X-ray photograph of
DNA were shown to Watson in 1952 by one of her colleagues at King's,
Maurice Wilkins. The photograph clearly showed that DNA had a coiled
structure. This was the information that Crick and Watson needed. They were
then able to make an accurate model of DNA.

The scientific magazine, *Nature*, published three papers in 1953 about the
structure of DNA. One paper was by Crick and Watson, one by Wilkins and
one by Rosalind Franklin.

In 1962, Crick and Watson and Wilkins were awarded the Nobel Prize for
their work on DNA. In a book in 1965, Watson gave no credit to Rosalind
Franklin for her part in the discovery.

Kathleen Lonsdale (1903-1971)
became the first woman
to be elected to the
Royal Society in 1957.

Her studies in X-ray
crystallography included
the structure of
benzene and diamonds
found in meteorites.

Indeed, these diamonds
were called Lonsdaleite.

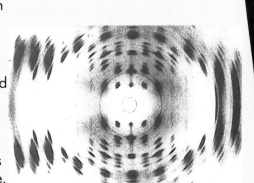

In 1964, Dorothy Hodgkin was awarded the Nobel Prize for her work in
X-ray crystallography. Her work included the structure of a protein, penicillin
and vitamin B_{12}:

(a) Name three famous women X-ray crystallographers.

(b) What was Rosalind Franklin's contribution to the structure of DNA?

(c) In science, there have often been disagreements about who had done
what in making a discovery. Find another example in this book.

MELTING, DISSOLVING AND DIFFUSION

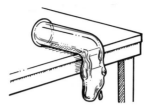  

Melting When a solid is melted the regular arrangement of particles breaks down and the particles become freer and able to move. This is shown by reference to the diagram of particles on page 73.

Dissolving If a beaker is completely filled with water and salt is then slowly added to the water, the salt dissolves. The water does not overflow from the beaker, however. The salt is made up of a giant structure of ions. When the salt dissolves the structure of ions breaks up and the ions fill spaces between the water particles.

Diffusion If you open a bottle of perfume in a room, the smell of the perfume soon spreads throughout the room. This can be understood, if the perfume is made up from millions of tiny particles which can move around the room. Diffusion is the movement of a gas to fill any space in which it is put and can be demonstrated in the laboratory by putting a gas jar filled with air above a gas jar filled with heavier, red-brown bromine vapour.

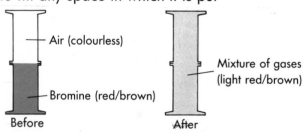

After a few minutes, the contents of the two gas jars look the same. The bromine particles have spread out evenly into both gas jars. This movement of particles is called **diffusion**.

Diffusion takes place in liquids but more slowly. This is because the particles in a liquid are moving slower than in the gas. If a purple crystal of potassium permanganate is dropped into a beaker of water, diffusion takes place. After some hours the whole solution is a pale pink colour. One small crystal of potassium permanganate must contain enough small particles to spread out and fill all of the water.

Can you think of other examples of melting, dissolving and diffusion?

UNIT 9
EARTH AND ATMOSPHERE

You should already know that weather conditions change and that some of these changes depend upon the season of the year. Weather is important in the life of most people. You will probably have recorded weather information over a short period of time and should know some of the common symbols used in newspapers and on television to represent weather conditions.

AIR PRESSURE AND WEATHER

The Earth is surrounded with a thick layer of gases called the **atmosphere**. Changes which take place in the atmosphere are called **weather** and the study of weather is called **meteorology**.

Km
300 -

200 -

100 -
Ionosphere
Mesophere   Stratospause
Stratosphere   Tropopause   Ozone layer
0   Troposhere

Earth's surface

The layer of air above the Earth is about 80 kilometres high. The weight of all this air pressing down is called **air pressure**. We do not notice this because it is always there. Air pressure is measured is millibars (mb) and an average value is about 1016 mb at sea level.

However, air pressure is constantly changing and is measured using a **barometer**.

The diagram shows two sorts of barometer.

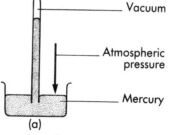

Vacuum

Atmospheric pressure

Mercury

(a)

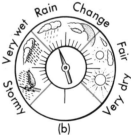

Very wet Rain Change Fair Very Dry Stormy

(b)

In (a) the air pressure supports a column of mercury about 76 cm high. As the air pressure changes, the height of the mercury column will change. Diagram (b) shows an aneroid barometer. This records air pressure on a scale.

When the air pressure rises above the average value, we have a **high pressure** area called an **anticyclone**. Areas of lower than usual air pressure are called **depressions**. They are shown on weather maps by areas marked 'HIGH' and 'LOW'.

When there is an anticyclone over the British Isles calm weather can be expected. Winds will be light and generally clockwise, slightly away from the centre of the anticyclone. In summer, this leads to cool misty mornings followed by hot, clear sunny afternoons. In winter, mornings can be cold, frosty and foggy, followed by cold, clear sunny afternoons.

With the depression, the winds blow in an anticlockwise direction slightly towards its centre. What do you think the weather will be like?

Many of the winds over the British Isles blow from the south-west towards the north-east. These winds are called **prevailing winds**. Anticyclones, depressions and other weather systems tend to move from south-west to north-east.

Here are two weather maps, one day apart.

On the weather maps, the lines join up places where air pressure is the same.

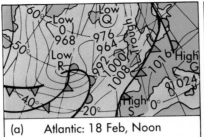

(a) Atlantic: 18 Feb, Noon

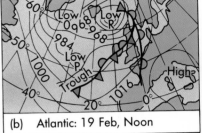

(b) Atlantic: 19 Feb, Noon

In (a), one line passing north of Scotland joins up places where the air pressure is 984 mb. These lines are called **isobars**. Generally, winds are stronger when isobars are close together. Wind is air, moving because of differences in air pressure.

Did you know?

Measuring windspeed

Sir Francis Beaufort (1774-1857) was born in Ireland. He joined the Royal Navy as a young man and served in it for over 20 years. In 1806 he put forward the idea of the Beaufort scale to measure wind speed. In those days the idea was to state the amount of sail a ship should carry at each wind speed. The scale was officially adopted by the Royal Navy about 1850. In 1829, he became hydrographer to the Royal Navy. A hydrographer is someone who studies water on the Earth's surface such as in oceans, rivers and lakes. The Beaufort scale is:

Beaufort scale number	Effects		Speed in knots (1 knot = 1.15 mph)	
0	calm	smoke rises vertically	0	
1	light air	smoke drifts in wind	1-3	
2	light breeze	wind felt on face, leaves rustle, weather vane moves	4-6	
3	gentle breeze	light flag blows, leaves and small twigs move	7-10	
4	moderate breeze	small branches move, dust and loose paper blows	11-16	
5	fresh breeze	small trees sway	17-21	
6	strong breeze	large branches sway	22-27	
7	moderate gale	hard to walk into wind, trees start to sway	28-33	
8	fresh gale	very hard to walk into wind, twigs break off trees	34-40	
9	strong gale	structural damage probable	41-47	
10	whole gale	trees uprooted, serious damage to buildings	48-55	

[continued on next page]

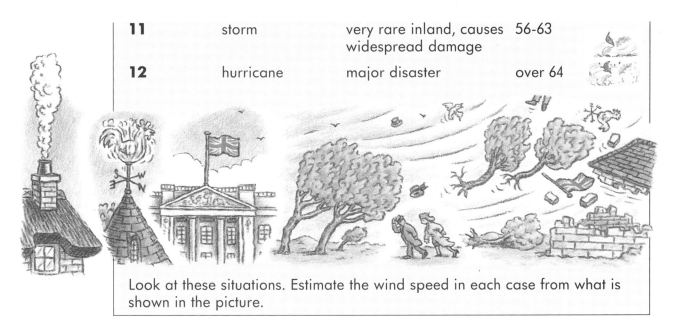

| 11 | storm | very rare inland, causes widespread damage | 56-63 |
| 12 | hurricane | major disaster | over 64 |

Look at these situations. Estimate the wind speed in each case from what is shown in the picture.

The amount of water vapour in the atmosphere is very variable. When water evaporates the air becomes saturated with water vapour. As the humid air rises and cools, clouds are formed. There are different types of cloud. These are shown in table 1.

Cloud type	Height	Description
cirrus	very high	wispy white threads
cumulus	medium	white 'cotton wool'-like
nimbus	medium	dark grey
stratus	low	continuous sheet of low cloud

Table 1

Can you identify the cloud types from these pictures?

When the water vapour condenses in a cloud, rain is formed. Under cold conditions, water vapour condensing can cause hail or snow.

There are distinct weather patterns over the Earth. Convection currents, that rise in the tropics and sink at the poles, produce circulating air currents or winds. This simple pattern is confused by the spacing of land masses and the sea and by the rotation of the Earth.

When cold air moves underneath an area of warm air, a **cold front** is set up. A **warm front** is formed when warm air moves above cold air. An **occluded front** is formed when a cold front catches up a warm front. Table 2 shows the symbols for these fronts. As a warm front passes, there will be increasing wind and cloud, light rain and a fall in air pressure. Heavy rain, very strong wind and a drop in temperature accompany the passing of a cold front.

Type of front   Symbol

(a) Cold front

(b) Warm front

(c) Occluded front

Table 2

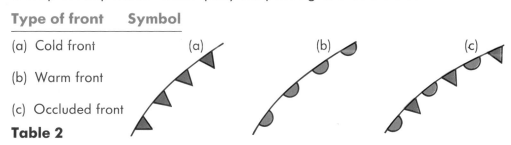

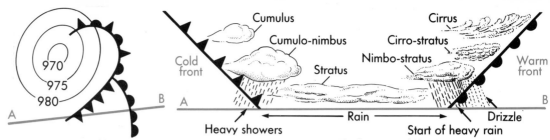

The illustration shows the clouds and weather along a line AB.

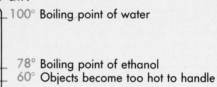
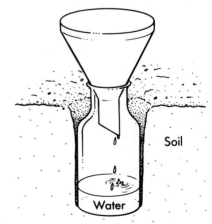
ACTION!

Measuring rainfall

It is easy to make a simple rainfall gauge. All you need is a funnel and a flask to collect the water. This is then sunk into the ground.
The amount of water is collected and measured in a measuring cylinder at the same time each day.

The table shows the results obtained in two weeks in April.

Day of the week	Rainfall in mm	
	Week 1	Week 2
Monday	2	0
Tuesday	2.5	1
Wednesday	3.5	9
Thursday	1.5	3
Friday	0	8
Saturday	0.5	0
Sunday	4	0

Soil

Water

(a) Plot these results on a suitable block graph.

(b) In which week was there the greatest rainfall?

(c) What was the average rainfall each day during week 1?

(d) What was the average rainfall each day during week 2?

(e) On the basis of these results only, is it fair to conclude that Wednesday is the wettest day **each** week?

(f) What else would you do to confirm you answer to part (e)?

ACTION!

Under pressure

The diagram is a weather map of Great Britain.

Use this map to answer the following questions:

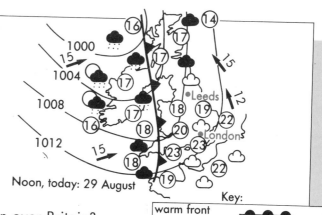

Noon, today: 29 August

Key:
warm front	●●●
cold front	▲▲▲
occluded front	▲●▲
temperature (°C) — black circles	
wind speed (mph) — arrows	
pressure (millibars) — isobars	

(a) Is there a HIGH or LOW pressure area to the north-west of Scotland? Explain your answer.

(b) What kind of front is shown over Britain?

(c) What is the temperature in London?

(d) Estimate the wind speed and direction in London.

(e) Describe the weather in Leeds. How would you expect the weather to change if the front continues to move in an easterly direction?

THE WATER CYCLE

Look at the water cycle diagram.

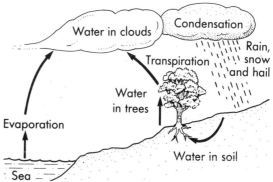

This shows how the water on the Earth passes through different stages and finishes up where it begins.

There is a natural cycle involving evaporation and condensation.

This natural cycle is very similar to the process of distillation which we use in the laboratory (*see* page 55).

Water evaporates from seas, rivers and lakes and the water vapour rises into the atmosphere. When the water cools it forms clouds. The clouds travel on air currents. When the water vapour condenses, the droplets of water fall as rain. Much of the water which falls on the ground eventually finds its way back into the rivers, lakes or the sea.

Water is essential for the life of humans and other animals. Animals lose as much water as they take in because they lose water when they sweat, urinate and breathe out. Make a list of the ways in which water is vital in your life.

THE EARTH'S CRUST

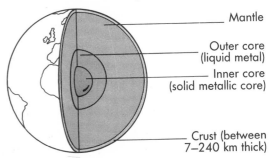

Look at the structure of the Earth.

The thin layer on the surface of the Earth is called the **crust**.

This consists of large plates of rock which are floating on the molten mantle.

- Mantle
- Outer core (liquid metal)
- Inner core (solid metallic core)
- Crust (between 7–240 km thick)

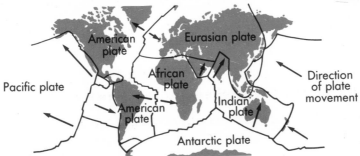

The map shows how the plates are distributed over the surface of the earth. The study of these plates and their movement is called **plate tectonics**. The movement of these plates is extremely slow, perhaps 1-2 cm per year. When plates collide important changes occur. It is thought that the Himalayan mountains were formed from the collision of the Indian and Eurasian plates.

In some places molten rock or magma, which is below the plates, escapes from the Earth as lava in volcanic eruptions.

Look at the areas where earthquake and volcanic activity occur.

Notice that activity occurs where the edges of plates come into contact.

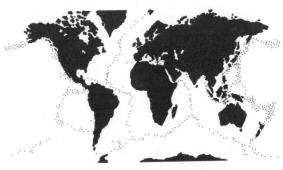

Earthquakes

Scientists think that earthquakes occur when the plates of the earth rub together. The study of earthquakes is called **seismology**. The strength of an earthquake is measured on the Richter scale:

Richter scale reading	Comment
8.9	highest recorded earthquake
8	
7	
6	severe earthquake
5	
4	
3	
2	earth tremor
1	

On this scale, an earthquake measuring 7 is ten times stronger than one measuring 6, and one hundred times stronger than one measuring 5.

There are about 10 000 earthquakes each year but, thankfully, only about ten are severe enough to cause deaths.

When an earthquake occurs, cracks appear in the ground.

These may be over 5 m wide.

Buildings collapse, gas pipes split and electricity supplies are disrupted.

Often landslides, floods and tidal waves accompany earthquakes.

In October 1989, there was a large earthquake in San Francisco. The earthquake registered 6.9 on the Richter scale. Over 270 people died and the damage was estimated at about £2.5 billion. However, the damage was not as severe as the earthquake in 1906. In 1906, most of the damage was caused by fires which followed the earthquake from broken gas mains.

The earthquakes in San Francisco were caused by movement of the Pacific and American plates, along what is called the San Andreas Fault.

There are 300 small earthquakes along this fault each year.

Modern technology has enabled builders to build tall buildings which are not destroyed by earthquakes.

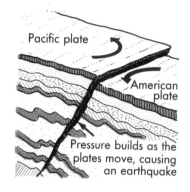

Pressure builds as the plates move, causing an earthquake

Buildings are at their most vulnerable when their foundations are of similar dimensions to the tremor wavelengths

Using data from sensors, buildings can be designed to ride earthquakes by making their foundations larger than the wavelengths of the tremors likely to hit them

How else do you think that these skyscrapers might be built to stop them from falling down in an earthquake?

Volcanoes

The pushing of two plates together causes an earthquake. It is also possible for one of the plates to force the edge of another plate downwards. The friction of the two plates rubbing on each other creates heat. This heat melts some of the rock. The molten rock travels along any cracks and lines of weakness in the Earth's crust. These cracks may continue up to the surface so that molten rock runs through them forming a volcano.

A volcano erupts with a huge explosion because water is turned into steam near the surface.

Other gases are also boiled out of the material coming from deep in the Earth.

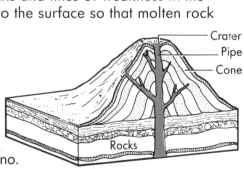

The diagram shows a cross section of a volcano.

When a volcano is first formed, the exploding gases force their way to the surface and make a large crater. Molten rock thrown out of the crater is called lava.

There are about 500 active volcanoes on the Earth. There are, however, more active volcanoes under the oceans. Only a small number of volcanoes erupt each year. A dormant volcano is one where the lava has solidified in the crater.

It may still erupt, however, even with a large explosion.

Volcanoes can be thought of as safety valves for the Earth. If the pressure builds up in the Earth, the pressure can be released by an eruption.

Volcanic eruptions can also affect the weather. Find out how this may happen!

Rocks

When considering the rocks of the Earth, it is difficult to understand the huge lengths of time involved. It has taken 4500 million years for the Earth to be at the stage it has reached today. You get a better idea if we use a time-chart of the history of the Earth. It we take a year to represent the life of the Earth, each day represents 12.3 million years.

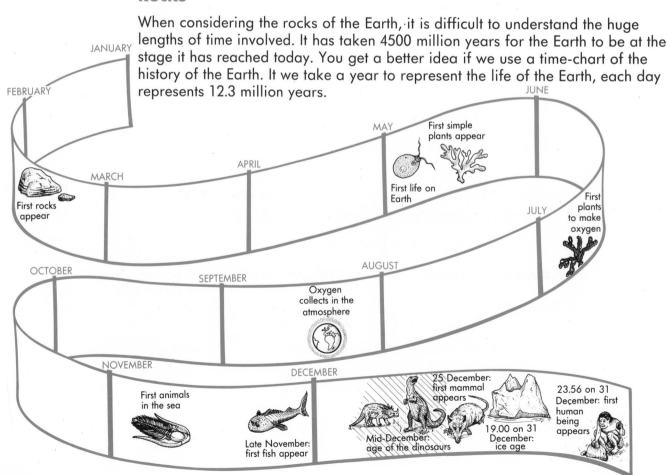

JANUARY

FEBRUARY

MARCH — First rocks appear

APRIL

MAY — First life on Earth

First simple plants appear

JUNE

JULY — First plants to make oxygen

AUGUST — Oxygen collects in the atmosphere

SEPTEMBER

OCTOBER

NOVEMBER — First animals in the sea

Late November: first fish appear

DECEMBER — Mid-December: age of the dinosaurs

25 December: first mammal appears

19.00 on 31 December: ice age

23.56 on 31 December: first human being appears

ACTION!

There are different types of rock in the Earth.

1 Sedimentary rocks The rocks in the Earth are eroded or broken down in many different ways. The result is a sediment made up of many different fragments. When these layers are compressed over millions of years sedimentary rocks are produced. Chalk and limestone are sedimentary rocks.

2 Igneous rocks Igneous rocks are formed when molten magma from inside the Earth is cooled. The size of the crystals produced depends upon the rate of cooling. Granite and basalt are different types of igneous rocks.

3 Metamorphic rocks Very high temperatures and pressures convert sedimentary rocks into metamorphic rocks. Marble is a metamorphic rock formed from limestone. Slate is a metamorphic rock formed from mud.

1

2

3

Rocks are continually being broken down or weathered.

However, the total amount of rock in the Earth's crust does not decrease.

New rocks are constantly being formed in the **rock cycle**, as shown opposite.

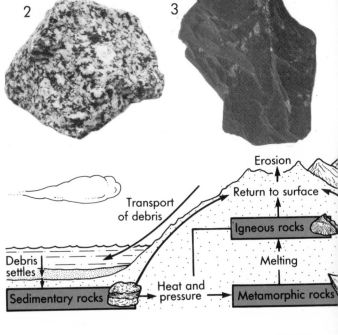

Erosion

Transport of debris

Return to surface

Igneous rocks

Debris settles

Melting

Sedimentary rocks

Heat and pressure

Metamorphic rocks

ACTION!

Ages of rock

Draw the diagram and complete it by adding labels.

The labels you can add are: erosion; heat and pressure; cooling; compacting; melting.

They can be used more than once.

Magma (molten rock)

2

1

Igneous rock

Metamorphic rocks

6

5

3

7

Sediment

4

Sedimentary rocks

Weathering and landscaping

Modern buildings are often made of brick or concrete rather than of natural stone. Where natural stone is used it is often sandstone or limestone. Both sandstone and limestone are easily weathered (or **eroded**). Limestone is badly eroded by acid rain. How do you think this process might occur? Erosion of stone can be seen in statues and church buildings.

ACTION!

Fossilised rocks

Marble, limestone and chalk are all forms of calcium carbonate. Why may fossils be found in limestone and chalk but not in marble?

Erosion is frequently caused by the effects of water.

Water expands when it freezes. When water gets into a crack in a rock, and then freezes, the crack is opened.

Can you think why freezing makes the cracks larger?

Over several winters, this can cause pieces of rock to fall off.

A pile of pieces of rock at the bottom of a rock face is called a **scree**.

ACTION!

Rainfall can cause serious erosion by washing away the surface soil.
This becomes more serious when forests and other surface vegetation are
removed.
Soil is washed off slopes and deposited on flat land. Rivers can erode the
landscape by washing away rocks and soil, and depositing them downstream.

The sea is a powerful eroder.

It can attack cliffs creating caves,
arches and other shapes.

The wind can also shape rocks too.
The Sphinx in Egypt has been eroded by
sand carried in the wind.

Soil

The weathering of rocks breaks them down into very small pieces which become
part of what we call soil.

The size of the pieces (called grain size) varies in different soils.
A sandy soil drains well because it has large grains which have big spaces
between them. A clay soil soon becomes waterlogged because the spaces are
small and they soon fill up with water.

Sand → 2 mm Silt → 0.2 mm Clay

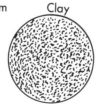

Magnified 100 times

Look at the soil profile
formed when soil is mixed
with water and allowed to settle.

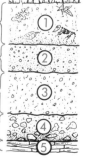

The organic horizon is dark
brown. It contains dead
plants and animals. ①

The topsoil. Rain water washes minerals to lower levels.
This process is called leaching. ②

Less humus found here but the minerals from 2 are
deposited. Iron may precipitate out, giving the soil a
yellow colour. ③

Small rock fragments are found here because the rock
is still breaking down. ④

Parent, bed rock ⑤

A fertile soil is good for gardening or farming. It should contain a lot of rotted vegetation called **humus** and be a **loam** soil. A loam soil contains grains of different sizes.

ACTION!

Water retention in soil

Samples of three soils were compared in an experiment. One soil was a sandy soil, another was a clay soil and the other a loam. Samples of the soils were placed in separate funnels placed in measuring cylinders. Equal volumes of water were added to each funnel.

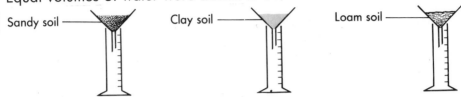

Sandy soil   Clay soil   Loam soil

(a) What preparation of each sample of soil should be carried out before the experiment?

(b) Complete the table by adding the following results in the correct places:

10 seconds; 60 seconds; 250 seconds; 0.5cm³; 30cm³; 50cm³

	Sandy soil	Clay soil	Loam soil
Time taken for first drop to fall into cylinder	_____	_____	_____
Volume of water collected after 15 minutes	_____	_____	_____

UNIT 10

FORCES

You should know by now that an object can be made to start moving, speed up, change direction, slow down or stop by pushing or pulling.

FORCES

Forces are responsible for starting or stopping things or changing their shape. For example, a force is needed to push a car which has broken down and another force is needed to stop a car which has started to roll down a hill. Pushing and pulling are types of force.

Forces are measured in newtons (N) using a newtonmeter. These units are named after the famous scientist, Sir Isaac Newton.

A newtonmeter or forcemeter is a spring which extends when a force is applied.

The bigger the force the more the spring extends.

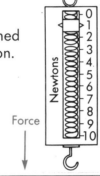

ACTION!

Pulley forces

In the diagram a string is stretched over a pulley between two fixed points.

The forcemeters show the force operating in each case.

The shaded area on the forcemeter is a measure of the force operating.

(a) What can you conclude from these results?

(b) What should you do to confirm these results?

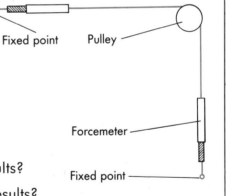

A force always acts in a definite direction.

Forces usually act in pairs.

For example, when you stand on the floor you exert a force on the floor.

The floor exerts an equal force upwards on to your feet which stops you sinking into the ground! Can you think of a different example?

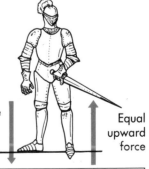

Downward force due to gravity

Equal upward force

ACTION!

Comparing strengths

Use sheets of writing paper to make bridges between two bricks (or blocks of wood). The drawing on page 90 shows two model bridges.

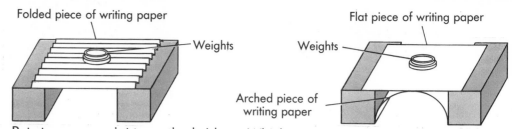

Folded piece of writing paper — Weights

Flat piece of writing paper — Weights

Arched piece of writing paper

Put stones or weights on the bridges. Which one supports the weights better?

Look at other bridges and try and understand how the bridge supports the downward forces.

Weight is a common force. A bag of granulated sugar has a **mass** of 1 kg. If it is hung on a newtonmeter, it is found that the force exerted is about 10 N. The force pulling down on the bag of sugar is caused by the force of gravity from the Earth. If the bag of sugar is taken to the moon, the mass of the sugar is still 1 kg but its weight is less than 2 N because the gravity on the moon is about six times weaker than the Earth. In outer space, where there is no gravity, the bag of sugar (still mass 1 kg) has no weight. Therefore it is weightless.

ACTION!

Specific gravities

The Moon's gravity is about one sixth of that of the Earth.
An object of mass 120 kg is weighed on the Earth and on the Moon.

(a) What is the weight of the object (in N) on the Earth?

(b) What is the mass of the object on the Moon?

(c) What is the weight of the object (in N) on the Moon?

On the Earth, a mass of 1 kg exerts a downward force of 10 N. If the mass rests on a table, there must be a force of 10 N upwards. These two forces balance out.

ACTION!

Forced to weight

What would be the force downward exerted by masses of:
(a) 5 kg; (b) 50 kg; (c) 1 tonne (i.e. 1000 kg)?

What are the masses of objects with the weights given below:
(d) 50 N; (e) 500 N; (f) 10 000 N?

The downward acting force, due to gravity acting on the mass, is called weight. Weight depends upon the force of gravity.

Satellites orbiting the earth follow a circular path around the earth. The pull of the Earth due to gravity keeps the satellite in its circular orbit. Can you explain why the satellites stay in orbit?

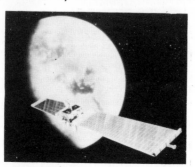

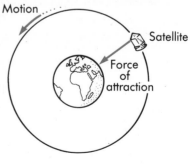

Motion — Satellite — Force of attraction

Sir Isaac Newton was born in 1642. Newton was the son of a Lincolnshire farmer. He was often unwell and extremely shy as a child. Since he was not making progress at school in Grantham, his mother took him away so that he could follow in his father's footsteps as a farmer. He did not enjoy the life of a farmer.

His uncle persuaded his mother to return him to school. He became a changed boy after he stood up for himself against the school bully. He went on to study at Trinity College at Cambridge University.

In 1665, the University was closed because of the plague. During this time, Isaac studied himself and mapped out some of the ideas which he was to follow later. At 27 he was made Professor of Mathematics at Cambridge. The theory of gravity, which he published in 1684, was probably his greatest contribution to science.

Despite being the first scientist knighted, and becoming President of the Royal Society, Newton was not liked by many people. He spent much of his later life in disputes with other scientists. One of these disputes was with the German philosopher Gottfried Liebniz. Independently, Liebniz and Newton had both developed the branch of mathematics called calculus. It is now believed that Newton had developed calculus first but Liebniz certainly published his work first. A major row developed over who had been first. Newton wrote articles, using his friends' names, to support his own claims. Liebniz was so annoyed that he made a complaint to the Royal Society about Newton. This was a very unwise thing to do as Newton was the president.

An enquiry was organized by the Royal Society. Newton ensured that the committee of enquiry contained a number of his friends and helped them to write the finished report. It should come as no surprise that the 'impartial' committee found in favour of Newton and accused Liebniz of copying Newton's work.

Newton is sometimes pictured sitting under an apple tree and being hit upon the head by an apple. People have said that because of this Newton came up with his theories of gravity. This is not true. It was common sense that if you drop an object it falls to the ground. Newton realized that a force of attraction exists between any two objects. The force depends upon the sizes of the two objects and how close together they are.

FLOATING AND SINKING

When an object floats in water, the object pushes away some of the water.

For an object which floats, the weight of water pushed away (called displaced water) equals the weight of the object.

The displaced water pushes up on the object.

This upward force is called **upthrust**.

If a mass which weights 10 N is lowered into water, it weighs only about 8 N.

The upthrust by the displaced water is 2 N.

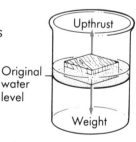

Original water level

Upthrust

Weight

The weight of this volume of water equals the mass of the object

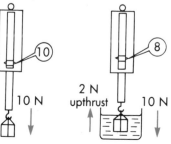

10 N

2 N upthrust

10 N

If the weight of an object is greater than its upthrust then the object sinks. If the weight of an object is equal to its upthrust then the object floats. Can you explain why icebergs exist?

This principle was first shown by Archimedes over 2200 years ago.

Using plasticine and a beaker of water, investigate the effects of changing the shape of an object!

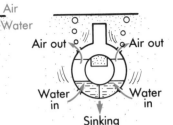

The submarine uses these principles of floating and sinking. In the design of the submarine there are large ballast tanks. When the submarine is floating on the water, the ballast tanks are full of air. When the submarine has to dive, the tanks are filled with sea water. The weight of the submarine is now greater than the upthrust and so the submarine sinks.

When the submarine surfaces again, water is pumped out of the ballast tanks. The weight of the submarine is reduced and the upthrust is sufficient to push it to the surface.

Did you know?

An elephant can weigh 11 tonnes and a whale can weigh 130 tonnes.

The skeleton is often the most important factor in determining how large an animal can grow.

A whale could not support itself on land.

The water provides upthrust to support the whale.

PRESSURE

Pressure is the spreading out of a force on a surface. A block of metal of mass 1 kg can stand on a surface in two different ways.

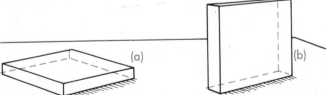

The block has dimensions 10 cm × 10 cm × 2 cm. The block exerts a force of 10 N downwards due to gravity. However, in (a) the force is exerted on an area of 100 cm² (10 cm × 10 cm) but in (b) it acts on an area of 20 cm² (2 × 10 cm).

Notice that in (a) the same force is spread over five times as much area as in (b).

Pressure is the force which acts on a single unit of area. In (a) the pressure is: 10 ÷ 100 = 0.1 N/cm²; and in (b) the pressure is 10 ÷ 20 = 0.5 N/cm²

A force, concentrated on a small area, will exert a much larger pressure than the same force acting on a large area. Notice that the units of pressure are N/cm² or N/m².

1 N/m² = 1 pascal (Pa). The pascal is the common unit of pressure. In Unit 9, when measuring air pressure, we used units of millibars (mb). 100 000 Pa,

(i.e. 100 000 N/m^2) is called a bar and a millibar is therefore 100 Pa.

In solids, pressure acts only in the direction of the force. Stiletto heels damage floors more than ordinary heels because, although the force is the same, the area in contact with the floor is much less. The pressure is therefore much greater.

ACTION!

Pressure points

Can you explain the ways in which pressure is involved in the cartoons?

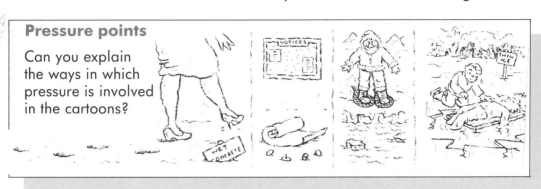

In liquids and gases the pressure acts in all directions.

If a force is applied to a liquid or a gas, it is transmitted through the liquid or gas to the walls of the container.

In a liquid, the pressure increases as the depth of the liquid increases.

This can be shown using a tin can as in the drawing.

The same is true for gases. The pressure of the Earth's atmosphere decreases as the height above the Earth's surface increases. The atmosphere can be thought of as a column of air 10 km high. This column has weight and provides a downward force.

The mercury barometer on page 77 can be used to measure the pressure of air. The pressure of air is balanced by the pressure of a column of mercury about 76 cm high. As the air pressure changes, so the height of the mercury column will change slightly.

Look at the **manometer** used to measure gas pressure.

The gas pressure forces some of the liquid from the left-hand side of the manometer to the right.

The difference in levels is a measure of the pressure of the gas.

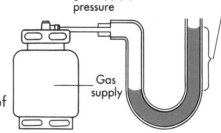

The difference in height indicates the extra gas supply pressure over the atmospheric pressure

Gas supply

In a car's braking system, the pressure applied to the brake-lever in the car is transmitted to the brake on the wheels by liquid.

FORCING ISSUES

Forces can change:

1 The shape or size of an object.
2 The direction of movement of an object.
3 The speed of an object.

A force can change the shape or size of an object. The force may change the shape of an object either permanently or temporarily. If the change is temporary, it is described as an **elastic** change.

For example, a spring is extended when a force is applied.

When the force is removed the spring returns to its original shape.

However, if the spring is extended too far, it will not return to its original shape and can no longer be used as a spring.

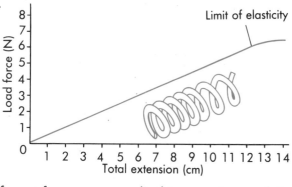

Look at the graph obtained when different forces are applied to a spring and the extension measured. From these results, it can be concluded that the extension of the spring is **directly proportional** to the force applied; providing the elastic limit has not been exceeded. Directly proportional means that the extension goes up as the force goes up. This statement is called **Hooke's Law**.

ACTION!

Read the following passage and use it to answer the question.

CHASSIS ALLOYS

Car bodies are usually made of sheets of steel. These sheets of steel are relatively inexpensive and can be pressed to give the required shapes. The steel is strong and does not easily change its shape. Steel, however, rusts readily and a great deal of money is spent preventing it from rusting.

Alternative materials are aluminium and glass fibre. Both do not rust and are less dense than steel. Aluminium is much more expensive than steel and less strong. Glass fibre is very light but not very strong. In even a slight accident the glass fibre would shatter.

'Newprod plc' is a company specializing in making new materials for industry. They are looking to produce a new material suitable for car bodies. They are hoping to interest a new Japanese car factory being built in England. They are looking at metals, polymers and ceramic materials for making these products.

What do you think they should look for in the ideal material?

Did you know?

Robert Hooke (1635-1703)

Robert Hooke was born at Freshwater, on the Isle of Wight, in 1635. At the age of twenty, he was employed by Robert Boyle (who used his scientific skill to make an air pump). In 1662, he was appointed curator of experiments to the Royal Society. He was elected a Fellow of the Royal Society in 1663.

Hooke's scientific achievements would probably be more significant if they had been less varied. He started a great deal but perfected little. He produced a simple form of the theory of light. He was the first to suggest that the movement of sun, moon, stars and planets could be worked out using the laws of mechanics. He devised a simple form of wheel barometer. He also designed a system, using a hairspring, for a balance, which was soon adapted by Thomas Tompion to operate a watch.

Hooke is remembered especially for Hooke's law (which can be applied to springs, rubber bands and even metal beams). This says that the extension of an object is proportional to the force applied to it. If a 50 N spring causes an extension of 1 cm, then a 100 N force will cause an extension of 2 cm.

Robert Hooke's scientific studies were very varied. Scientists have become very specialized nowadays, often only studying one tiny aspect of one branch of science.

(a) Why do you think scientists today have become more specialist than scientists were in the 17th century?

(b) What are the advantages and disadvantages of being specialized?

Forces can also squeeze or compress an object. Can you think of an example of this effect?

A force applied to the end of a spanner can turn a tight nut. The turning effect depends upon both the size of the force and the length of the spanner. The turning effect of a force is called a **moment** or **torque**. The moment of a force is calculated using the formula:

$$\frac{\text{moment}}{\text{(in Nm)}} = \frac{\text{force}}{\text{(in N)}} \times \frac{\text{distance from force}}{\text{to turning point (in m)}}$$

The turning point is called the **pivot** or **fulcrum**.

If a nut is very tight, a spanner with a long handle is much better at loosening it. The maximum force that can be applied is the same as that of a shorter spanner. However, the greater distance from the force to the turning point makes the moment much greater.

When a see-saw is balanced, the clockwise and anticlockwise moments are equal.

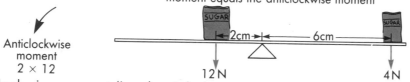

A balance occurs when the clockwise moment equals the anticlockwise moment

Anticlockwise moment 2 × 12

12 N   2cm   6cm   4 N

Clockwise moment 6 × 4

The clockwise moment (i.e. the right-hand side) is 6 × 4 = 24 Nm.
The anticlockwise moment (i.e. the left-hand side) is 2 × 12 = 24 Nm.

Therefore, the see-saw is balanced. If the 4 N force is moved 7 m from the fulcrum on the right-hand side, the clockwise moment becomes 4 × 7 = 28 Nm and the right-hand side will go down.

Many household items rely on turning moment mechanisms such as crowbars, wheelbarrows, nut crackers, scissors etc. Some of these are shown below.

Can you explain how they work?

wing boat

Nutcrackers

Scissors

Sugar tongs

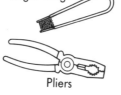

Pliers

Wheelbarrow

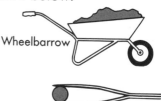

Tweezers

Speed involves moving from place to place. It gives a distance measured in an amount of time. It may be measured in a variety of units such as miles per hour (mph), kilometres per second (km/s) and metres per second (m/s) etc. Thus:

speed = distance ÷ time

The terms speed and **velocity** are often confused.

Velocity is speed in a certain direction.

Speed has no specified direction.

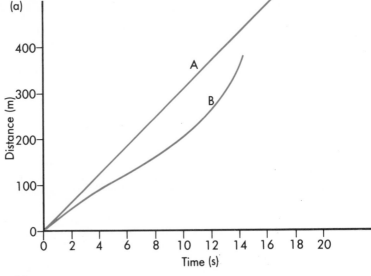

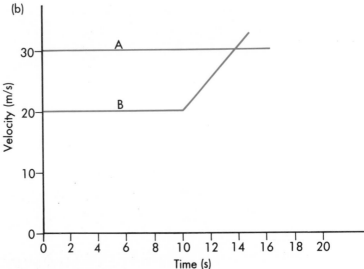

Graph (a) shows distance (on the *y* axis) against time for a car moving at a constant velocity of 30 m/s. This is labelled A. You will notice that the distance travelled each second is the same (30 m). In graph (b), the same car is shown in the graph labelled A. This is a graph of velocity against time.
The graph is horizontal (i.e. parallel to the *x* axis) as the speed is not changing.

Another car, B, is shown in the two graphs. You will notice that this car is not moving at a constant velocity throughout. After 10 seconds the car speeds up or **accelerates** so that every second it is travelling faster than before. Acceleration is measured in units of m/s^2 (i.e. metres per second per second). It is the average change of speed per second. Acceleration can be calculated using the equation:

$$\text{acceleration (m/s}^2) = \frac{\text{increase in velocity (m/s)}}{\text{time for the increase in velocity (s)}}$$

Slowing down is called **deceleration**. It has the same units as acceleration but, whereas acceleration is positive, deceleration is negative.

An object stays at rest (i.e. not moving), or moves at a constant speed in a straight line, unless a force acts upon it.

ACTION!

Mean velocity
What is the average speed of a car which travels the 160 miles from Stoke-on-Trent to London in 4 hours?

Travelling speeds

A car is travelling at 30 m/s. The driver then accelerates to 72 m/s in 7 seconds.

(a) What is the acceleration?

(b) The driver then has to brake suddenly and come to rest in 3 seconds. Calculate the deceleration.

(c) Explain why, without using a seat belt, the driver might hit the windscreen during a rapid deceleration.

Putting into force

Look at how forces act on a passenger in a car when the car is:
  (i) travelling at constant speed; (ii) rapidly accelerating; (iii) rapidly braking.

(a) Describe the forces that are acting in each case.

(b) The report on the Kegworth air crash in January 1989, recommended that it would be safer if all aircraft seats faced backwards instead of forwards. Explain why this recommendation was made.

Forces can make stationary objects start to move or alter the speed and/or the direction of a moving object.

force (N) = mass (kg) × acceleration (m/s²)

The equation shows how an object can be accelerated by applying a force. Acceleration depends upon the force applied (in N) and the mass of the object (in kg).

Mass acceleration

Calculate the acceleration of a 100 kg mass caused by a force of 300 N.

When an object is rolling at a steady speed, along a level surface, it will come to rest. This is because of a force of friction which resists motion. When a car is rolling, there is friction between the tyres and the road.

Friction is necessary to keep the car safely on the road.

Why do you think this is true?

Putting the brakes on

The table below is similar to one in the Highway Code and shows the overall stopping distances at different car speeds.

Speed of car (m/s)	10	15	20	25	30
Thinking distance (m)	6	9		15	18
Braking distance (m)	8	18		50	72
Overall stopping distance (m)	14	27		65	90

Table

The stopping distance is the sum of the thinking distance and the braking distance.

(a) What do you think is meant by the term 'thinking distance'?

(b) How does the thinking distance depend upon the speed of the car?

(c) What thinking distance corresponds to a speed of 20 m/s?

(d) Giving your reasons, suggest the safe minimum distance that a car should be driven behind another, if both are travelling at 15 m/s.

(e) Look at the graph of the braking distance against the (speed of car)^2. From the graph find the stopping distance for a car travelling at 20 m/s.

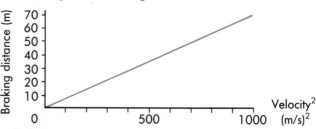

(f) Why is the graph a straight line?

(g) Why is the braking distance greater on wet roads?

(h) A car could be stopped by running it up a slope. If the car was travelling at 20 m/s, to what maximum vertical height h could it go before coming to rest? (Assume that friction can be ignored.)

PROJECT

1 Designing an all-purpose vehicle Design an all-purpose vehicle which will travel across any surface. It must not sink into sand or snow and must move smoothly on the surfaces.

Write about any special features which you would include in your vehicle as labels for your design sketch.

2 Skating on thin ice A young boy called Sam is skating on a frozen pond. The ice breaks and he falls into the water. How would you attempt to rescue him without falling into the water yourself? Explain in writing the scientific reasons for your course of action.

3 Coming to a standstill Friction is a force which opposes movement. When a block of wood is sliding down a slope a force of friction acts on the wood trying to prevent its movement. The size of the frictional force depends upon the materials used.

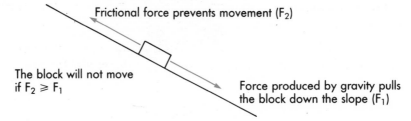

(a) Find out what you can about friction. Give examples of surfaces which have low friction.

(b) How can the frictional force between surfaces be reduced?

(c) When the brakes are used on a car or bicycle, friction is used to bring the car to a halt. What form of energy is produced when brakes are used?

UNIT 11

ELECTRICITY AND MAGNETISM

You should know that electricity has to be used with great care. When using electrical appliances you should follow sensible safety rules. You should also know that magnets attract certain materials such as iron and steel but not others like wood and aluminium. Two magnets, when brought together in a certain way, will repel each other.

CONDUCTORS AND INSULATORS

Look at the poster produced by the Electricity Council. It warns of the dangers of using a carbon fibre fishing rod close to overhead power cables.

If the fishing rod touches the cables electricity will pass through the carbon fibre and could easily kill the fisherman.

The fishing rod is said to **conduct** electricity and carbon fibre is called a **conductor**.

In 1984 a boy was killed flying a kite with a metal wire.

The kite became entangled with overhead power cables.

Metals behave in the same way as carbon fibre: they conduct electricity. If his kite had had a nylon line he would not have been killed. Nylon does not allow electricity to pass through it. It is called an **insulator**. Plastics, rubber, glass and most ceramic materials are good insulators. The poster shows ceramic discs on the pylons. These keep the cables away from the metal parts of the structure. The overhead cables are not usually insulated because the air around them is a good insulator. Can you think of other examples of conductors and insulators that you might use in everyday life? Write down three examples of each.

DON'T USE CARBON FIBRE RODS OR POLES NEAR OVERHEAD ELECTRIC POWER LINES IT COULD BE FATAL

There are some substances which conduct electricity poorly and only under certain conditions, such as when they are impure or when heated up. Silicon and germanium are examples. These substances are called **semiconductors**. They are used in transistors.

The apparatus below can be used to show whether a material conducts electricity. What happens if the material conducts electricity?

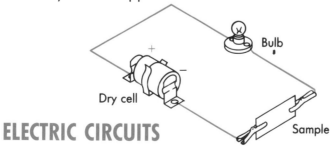

Material	Does the bulb light?
copper	✓
wood	✗
plastic	✗
iron	✓
lead	✓
graphite	✓

ELECTRIC CIRCUITS

There is no electricity without a source of energy. One convenient source of energy is a dry cell. A series of dry cells is called a **battery**. The energy is made

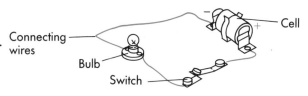

in the cell by chemical reaction between the substances in the cell.

When the reaction is completed the cell stops working.

The cell acts as a pump which keeps electricity flowing from one end of the cell to the other through the wire. Look at a simple electrical circuit above. When the switch is closed, the cell pumps electricity through the wire. The flow of electricity is detected by the bulb lighting up. Find out how lightbulbs work.

Imagine that the circuit above has two cells. Together, they have a larger **voltage**, they pump electricity more strongly and the same bulb glows brighter. No electricity will flow until the switch is closed because electricity only passes through a complete circuit. Electricity does not pass through the air and so opening the switch breaks the circuit.

ACTION!

Explain why he disconnected the battery quickly.

Pupil thinks quickly

Following the crash of a helicopter at Rochester a sixth form pupil at Abbots School, Andrew Nelson, was commended for his quick thinking. Immediately he rushed to the helicopter, crawled inside the wreckage and disconnected the battery . . .

Pictures of electrical circuits do not have to be drawn every time. Symbols are used for different electrical components to draw **circuit diagrams**. Look at the symbols for some of the components used so far. The circuit diagram shows the apparatus set up in the diagram at the top of the page.

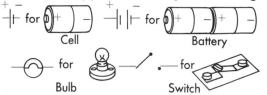

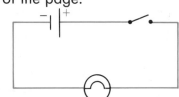

Here is another circuit diagram. This time, when the switch is closed, the bulb does not light up. The electricity always takes the easy path.

It is easier for the electricity to pass through the wire labelled X rather than the wire labelled Y.

The bulb provides a **resistance** to the passage of electricity.

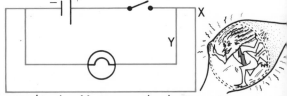

The passing of the electricity quickly through wire Y causes the battery to run out quickly. This wire will heat up and could lead to a fire. This rapid escape of electricity is called a **short circuit**.

Resistance and resistors

Anything which slows down the flow of electricity in a circuit is said to have a **resistance**. In the diagrams above the bulb is causing a resistance and is called a **resistor**. Look at the resistors, and the symbol used to represent a resistor, which are shown below.

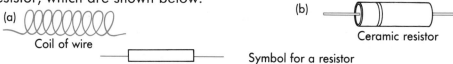

The resistance of each resistor is measured in units called ohms (symbol Ω). A simple resistor is a long length of wire made into a coil. The resistance of each resistor is fixed but it is possible to make resistors whose resistance can be varied. These are called **variable resistors**.

Look at these variable resistors.

The resistance can be altered by moving the slide or rotating the knob.

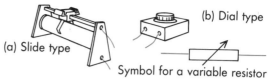

(a) Slide type

(b) Dial type

Symbol for a variable resistor

In homes, variable resistors are used to control the brightness of lights. These variable resistors are usually called dimmer switches.

Here is a simple circuit including a variable resistor.

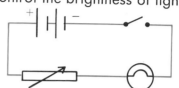

ACTION!

Completing a circuit

Draw a circuit diagram showing a circuit, including two bulbs, where one bulb can be switched on and off and the other bulb is on constantly but you can change its brightness.

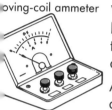
moving-coil ammeter

When the variable resistor is adjusted to have the maximum resistance, the battery may not be able to pump enough electricity through the circuit to light up the bulb. Very small electric currents can be measured using an instrument called an ammeter. Look at the ammeter and the symbol used to represent an ammeter.

An ammeter measures how many electric charges pass through it every second and uses units called amperes (usually shortened to amps or A)

Symbol for an ammeter

It is important to remember that in any circuit electricity flows at the same speed throughout. Four circuits, each with the ammeter in a different place are shown below. The reading on the ammeter is the same in each case.

Potential difference and voltage

If it was possible to measure the potential of the electric current before and after it enters a component in the circuit, it would give an idea of how much energy is being used up. This difference in potential (called **potential difference** or **p.d.**) is measured in volts (V).

A single dry cell battery is labelled 1.5 V meaning that there is a difference of 1.5 V between the two terminals.

In any circuit there is a difference in potential between any two points in the circuit.

An instrument which measures the potential difference is called a **voltmeter**.

Look at the voltmeter and the symbol used to represent a voltmeter.

The circuit shows the way a voltmeter can be used to measure the potential difference (p.d.) across a resistor.

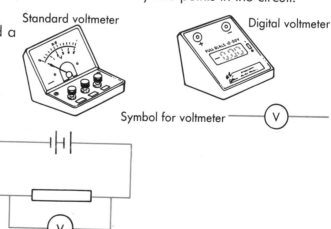

Standard voltmeter

Digital voltmeter

Symbol for voltmeter

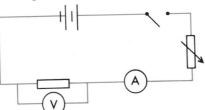

Realizing potential

Melanie set up the circuit below to measure the voltage across the resistor when different currents were passing through it.

Her results are in included in this table.

Current in amps	Potential difference in volts
0.5	1.0
1.5	2.8
2.0	4.0
3.0	6.0
4.0	8.0
5.0	10.0

(a) Name the pieces of apparatus labelled A and V.

(b) Why was the variable resistor included in the circuit?

(c) Look at the graph she drew of current against voltage. Four of her results are plotted on the graph. Plot the other two points and draw the best straight line through these points.

(d) Which result do you think is wrong? Explain the reason for your answer.

(e) What can you say about the relationship between current and voltage?

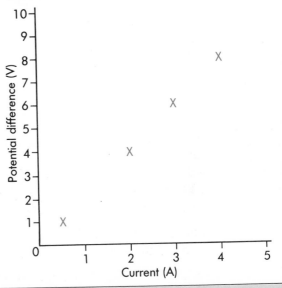

Series and parallel circuits

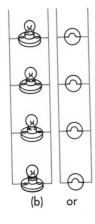

Bulbs can be connected together in a circuit in different ways. The diagram shows four bulbs connected in: (a) series; and (b) parallel.

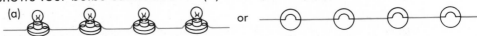

Christmas tree lights are usually connected in series. If one bulb stops working the circuit is broken and all bulbs go out. Also, each bulb removes some of the energy from the electric current. So the further along the circuit the bulb is, the less energy it will receive, and the less bright the bulb will shine. Outside garden lights are usually wired in parallel. This uses more wire and more electricity. However, all the bulbs are equally bright and if one bulb fails the other bulbs remain alight. Can you think why these circuits are the best way of wiring Christmas lights and garden lights respectively?

Direct current and alternating current

The electricity produced from the battery shown on page 99 is called **direct current (d.c.)**. The positive and negative terminals are fixed throughout and the current flows in one direction all of the time. The current is said to flow from the positive terminal to the negative.

Imagine that the connections on the battery were turned round over and over again. The direction of the current constantly changes. This is called **alternating current (a.c.)**. Household electricity is a.c. It changes direction of flow 50 times each second. It is said to have a frequency of 50 hertz (50 Hz). This frequency is set in the power station.

ACTION!

> ### Ring-mains: the wires and wherefores
>
> Look at the circuit diagram for a model of a circuit which exists in most houses.
>
>
>
> (a) Label the electrical components in the diagram.
>
> (b) What will happen if either of the switches is operated once?
>
> (c) Where in a house would you find this kind of switch and why is it useful?

STATIC ELECTRICITY

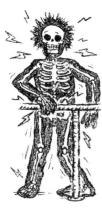

If you are walking through a department store and you touch a metal counter fitting, you can sometimes feel a small electric shock. It is not a system to stop shoplifters. There is no electricity passing through the metal fittings.

The electricity is stored up in you, probably on your clothes. It is called **static electricity**. When you touch the metal fitting, or any other metal object, the electricity flows through you. The escape route for the static electricity is called an **Earth connection**. When the current of electricity flows, you will feel the sensation of an electric shock.

Static electricity builds up on an insulator like a plastic material.

The insulator becomes **charged** with static electricity.

The insulator will either be positively charged or negatively charged.

Electricity charges can be detected with an electroscope.
A small piece of gold leaf is attached to a central metal rod. The metal rod passes through an insulated cap. The whole arrangement is surrounded by a glass jar to stop draughts affecting it.

The rod can receive a charge by either:

1 adding negative charges, which makes the electroscope negative

2 removing negative charges, which make the electroscope positive

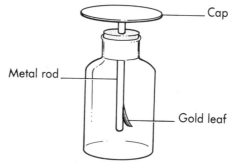

In either case, the charge spreads over the whole rod and the gold leaf. Since like charges repel, the leaf will be pushed out from the rod. If the rod is touched with the finger, the charge escapes and the leaf returns to its original position.

ELECTRICAL EFFECTS

There are three main effects of an electric current that is flowing. These are:
a magnetic effect; a heating effect and a chemical effect.

1 Magnetic effect When an electric current passes through a wire it produces a magnetic field around the wire.

2 Heating effect When an electric current passes through a wire the wire heats up. When a bulb lights, the very thin filament in the bulb gets white hot and gives out light.

In a circuit, a **fuse** is often included as a safety device. The fuse contains a wire which melts and breaks the circuit if it overheats for any reason.

3 Chemical effect Electricity can split up chemicals and cause chemical reactions to occur. This is called **electrolysis**. (*see* page 63)

ELECTROMAGNETISM

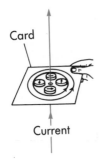

Card

Current

When an electric current flows through a fine wire (filament) in a bulb, it produces light and heat. When an electric current is passing through a wire, it also produces a magnetic field around the wire. This was first discovered in 1819 by Hans Christian Oersted.

If d.c. is used, the magnetic field will be in one direction only. The magnetic field is shown by arrows around the wire in the diagram.

The direction of the magnetic field can be found using the **right-hand grab rule**.

When the wire is gripped with the right hand with the thumb pointing in the direction of the current, the fingers show the direction of the magnetic field.

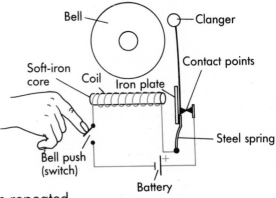

Current direction

Magnetic field direction

The magnetic field produced by a wire is relatively weak. A stronger field can be produced by winding the wire into a coil to produce a **solenoid**. The magnetic field can be made stronger still by putting a bar of soft iron in the middle of the coil. The magnetic field can also be made stronger by increasing the number of turns of wire in the coil or increasing the current flowing in the wire. The resulting magnet only has magnetic properties when the current is flowing and is called an **electromagnet**. The coil of wire behaves like a bar magnet. If a.c. is used, the magnetic field will alternate at the same rate as the current.

There are a number of everyday objects which rely upon electromagnetic effects:

1 Electric bell The diagram shows an electric bell. When the bell-push is pressed the circuit is made.

The electric current makes the coil into a magnet which attracts the iron armature. This causes the clanger to strike the bell and make a sound.

However, this also separates the contact points and cuts off the current. The clanger returns to its original position and the procedure is repeated.

Bell

Clanger

Contact points

Soft-iron core

Coil

Iron plate

Steel spring

Bell push (switch)

Battery

2 Electromagnetic relays *See* page 115.

3 Tape recorder Magnetic recording tape consists of a plastic tape coated with a magnetic material like iron oxide or chromium oxide. The tape stores up magnetic messages as it passes through the recording head.

Each message is caused by the turning on or off of the electromagnet in the recording head.

The electromagnet in the recording head is switched on and off by electric impulses from the microphone.

The tape produces a copy of the original sounds when the messages on it are passed by the playback head.

MAKING ELECTRICITY WITH A MAGNET

An electric current is made when a magnet is moved in and out of a coil of wire. The current is small and can be detected with a sensitive ammeter called a **galvanometer**.

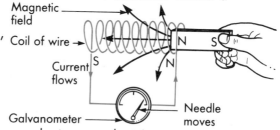

When the magnet is pushed into the coil, the needle on the galvanometer moves one way.

On removing the magnet, the needle moves the other way.

When a magnet is moved near a suitable conductor, an electric current is produced. This is called **electromagnetic induction**.

Generators

A generator converts kinetic energy into electrical energy. There are two types of generator:

1 a d.c. generator which produces d.c. electricity, (e.g. a bicycle dynamo).

2 an a.c. generator or alternator which produces a.c. electricity, (e.g. in a power station or a car).

In a bicycle dynamo, a magnet rotates near a coil of wire which is wound on a piece of soft iron.

A current is set up in the circuit which includes the coil of wire.

This makes it easier to rotate the magnet since rotating the coil would need some system for preventing the twisting of the wires.

The faster the magnet rotates the greater will be the current produced.

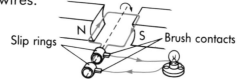

A simple a.c. generator or alternator is shown above. It has a rotating coil and a fixed magnet. As the coil rotates, a current is induced. The brush contacts enable the coil to rotate whilst maintaining contact with the rest of the electrical circuit. Three ways of increasing the current are:

1 spinning the coil faster   **2** using a stronger magnet
3 using more turns of wire on the coil

Electric motor

When a wire carrying an electric current passes through a magnetic field, a force acts on the wire and it moves. The direction of movement can be predicted using

Fleming's left-hand rule. This is the principle on which the simple d.c. electric motor is based. A motor turns electrical energy into kinetic energy.

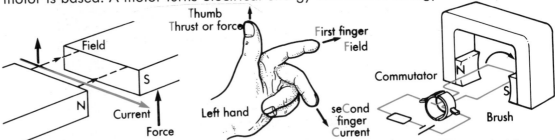

When the electric current passes through the coil, it produces a magnetic field. The interaction of this magnetic field and the field of the permanent magnet makes the coil move. The only possible movement is rotation and so the coil rotates. The motion is continuous and in one direction. The direction can be predicted by making use of Fleming's left-hand rule. If the direction of the current is reversed, the coil rotates in the opposite direction. The coil can be made to rotate faster by:

1 increasing the current
2 using a stronger magnet
3 using more turns of wire on the coil

The brushes are used to maintain electrical contact with the spinning coil. The commutator rotates and makes sure that whichever wire is nearer the north pole of the magnet, it always has the current moving in the same direction, so ensuring that the coil rotates in the same direction.

Electric motors used in everyday appliances such as electric drills, washing machines and food mixers usually have several coils wound onto one core. This is called an armature. Each of the coils has its own commutator. The additional coils make a stronger magnetic field and the resulting motor runs smoothly and with a more powerful turning effect.

ACTION!

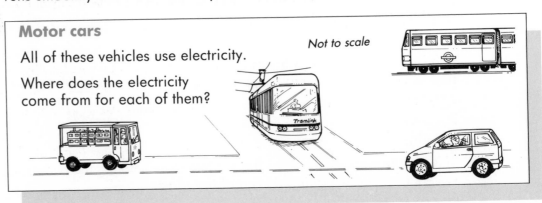

Motor cars

All of these vehicles use electricity.

Where does the electricity come from for each of them?

Not to scale

ELECTRICITY BILLS

Everyone has to pay for electricity that is used. The amount you have to pay depends upon:

1 how many appliances are being used
2 how long they are used for
3 how much electricity each appliance uses per second
   (the power rating of the appliance)
4 the cost of electricity

The power rating of an appliance is calculated by multiplying the current it takes (in A) and the voltage (in V). The units are watts (W) or kilowatts (kW):

power (in W) = current (in A) × voltage (in V)

The basic unit used for calculating the cost of electrical energy is the kilowatt-hour (kWh). This is the electricity used if 1000 W (or 1 kW) of electricity is used for 1 hour. Sometimes this is called 1 unit of electricity.

The amount of electricity used is found by reading the electricity meter.

Look at the sample electricity bill.

The difference between the present and previous readings is the amount of electricity used.

On top of the charge for the amount of electricity used, there is also a standing charge.

This covers administrative charges, etc.

HEB Electricity bill

Date: 5 April   From: 1 January
                 To: 1 April

Meter reading		Details	Amount
Present	**Previous**	DOMESTIC	
3045	2875		
		(VAT zero)	170 units @ 10p £17.00

Usually, a lower rate is available for electricity used during the night.

This is cheaper because the demand for electricity is less overnight and electricity can be generated more cheaply.

What would be the cost of using a three kilo-watt electricity fire for ten hours? (Assume electricity costs 10p per unit.)

quantity of electricity used = power rating (kW) × time (h)
$$= 3 \times 10 \text{ kWh}$$
$$= 30 \text{ units}$$

cost = number of units × cost per unit
$$= 30 \times 10 = £3.00$$

ACTION!

Reading your electricity meter

Read your electricity meter at home at the same time every day for three weeks. You can then work out how many units of electricity used by your family each day.

(a) Plot the results on a block graph.

(b) Is there any day of the week where you use more electricity than on a typical day? Can you explain why there is this difference?

(c) How many units of electricity did your family use over the three week period?

(d) What was the cost of the electricity used? Assume 1 unit of electricity costs 10p.)

(e) If you used electricity at the same rate throughout the year, what would be your charge for electricity for a whole year?

Watts the charge?

A household uses 720 units of electricity. The cost per unit is 10p and the standing charge is £14.

What is the cost shown on their bill?

The science of appliance

The diagram shows different electrical appliances and the time that each appliance takes to use one unit of electricity.

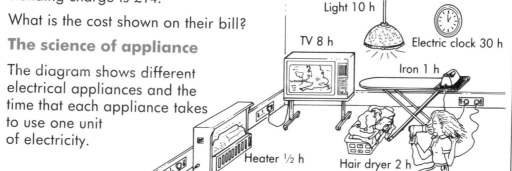

Light 10 h
TV 8 h
Electric clock 30 h
Iron 1 h
Heater ½ h
Hair dryer 2 h

(a) Arrange the appliances in order starting with the cheapest to run and finish with the most expensive.

(b) Work out how many units are required to run each appliance for one hour.

(c) Calculate the cost of:
    (i) watching television for four hours
    (ii) heating the room for five hours
    (iii) ironing for one hour
    (Assume 1 unit of electricity costs 10p and ignore standing charges.)

Using electricity safely

Did you know?

1 Always switch off electricity at the mains before attempting any electrical repair.

2 If in any doubt, leave repairs to an expert electrician.

3 Never allow electric wires of gadgets to come in contact with water or wet hands. Do not use mains electrical equipment in a bathroom.

4 Never make temporary repairs using makeshift materials and never use broken sockets, etc.

5 Never overload circuits by using adaptors.

6 Do not attempt to use appliances which use large amounts of electricity on a lighting circuit (e.g. washing machine).

ACTION!

Handle with care

These pictures show situations where electricity is not being used safely.

Explain the hazard in each picture.

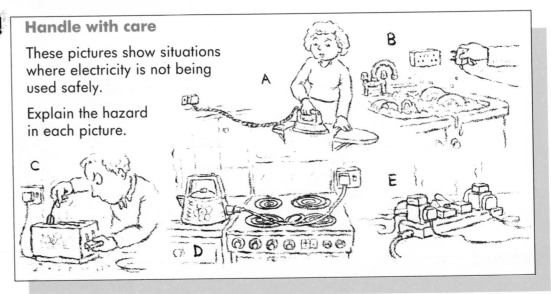

108

UNIT 12

ELECTRONICS AND MICROELECTRONICS

You should know that we can communicate rapidly over long distances in a number of ways. Some methods of communication allow two-way exchange of information, (e.g. telephone, fax, radio). Other methods, such as television and teletext, enable information to be received but not transmitted.

A range of everyday devices, including a computer, can be used to store information. Information can be processed and retrieved in various forms.

ELECTRONIC AND MICROELECTRONIC EQUIPMENT

A large number of electrical gadgets are used in daily life. People have different amounts of control over these different gadgets. For example, a table lamp has a simple on-off switch. It can either be on or off. You have much greater control of a television set. Apart from being able to turn it on and off, you can adjust the volume, the brightness and the colour of the picture.

Equipment that is simply switched on and off, is called **electrical** (e.g. the table lamp). Other equipment, such as the television set, contains special parts which enable more control. These parts include transistors, diodes, capacitors and resistors. These are called electronic components and the equipment is called **electronic**.

More recently we have combined several electronic components into integrated circuits which do the jobs of several components.

These miniaturized circuits are made possible by using slices of silicon called silicon chips.

The circuits are very small and the equipment is said to be **microelectronic**.

The photographs show how there has been a reduction in size of equipment in the past few years.

ACTION!

> **Techno-logs**
>
> Make lists of different electrical, electronic and microelectronic goods.

Switches

To understand microelectronic circuits, you need to know about switches. The diagram below shows a simple on/off switch. On the left-hand side, the switch is open (off) and on the right-hand side the switch is closed (on). When the switch is put in the open (off) position, it stays off.

When the switch is put in the closed position (on), it stays on.

You can use this kind of switch to operate a bell or light bulb.

Look at the electric bell circuit.

When the switch is closed, electric current flows and the bell starts to ring.

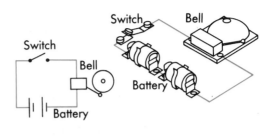

The flow of electricity is described as an electrical **signal**.

In this case, you decide by operating the switch, when you want to send the signal.

The signal operates the bell which is called the **output** device because it gives out the information as sound.

ACTION!

Setting alarm bells ringing

This fire-alarm system contains a bimetallic strip.

When a fire occurs, the strip bends and a contact is made.

Electricity flows through the circuit and the alarm bell rings.

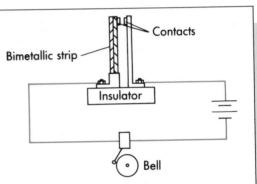

(a) If the wiring of the circuit burns out during the fire, what happens to the alarm? Is this a disadvantage of this type of alarm?

(b) Devise a fire-alarm system, based on three thermistors in different parts of the factory, such that a bell will ring if any of the thermistors record a rise in temperature.

A school bell can be rung manually by pressing a bell-push.

In microelectronic equipment, the **decision** is made by the microelectronic circuits in the silicon chips and not by a person. This could enable the bell to be rung at certain set times.

This can be summarized as:

Think again about the manual ringing of a bell. There are different decisions you can make. For instance: how long should the bell be rung; or how to stop the bell from ringing. Make a list of the other decisions that you would have to make.

The second decision is easy. You stop the bell ringing by opening the switch. How long the bell should be rung may depend upon information collected by **sensors**. These are **input** devices because they give you information. In the case of the school bell the sensors could be attached to a clock mechanism so the bells will ring at certain times.

At the time when the bell should ring a message is sent from the sensor.

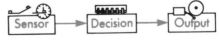

ELECTRONIC CONTROL SYSTEMS

There are two ways of showing the time on a wristwatch. These are shown below.

(a)

(b)

One method involves producing a small clockface. The second, minute and hour hands move round the face in a 'clockwise' direction. The mechanism makes the hands rotate. This is called an **analogue** display. Apart from clocks and watches, many instruments have an analogue display. Examples include many car speedometers, ammeters etc.

The second method involves a **digital** display. In electronics we frequently use digital systems. In a digital system only two possibilities exist:

logic level 1   i.e. **on**
logic level 0   i.e. **off**

Consider the simple circuit above, consisting of a battery, bulb and switch. There are two possibilities here:

Switch	Bulb
0 (off)	0 (off)
1 (on)	1 (on)

Table 1

Table 1 is called a **truth table**.

It summarizes the possibilities that exist in this simple circuit.

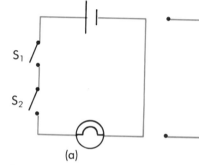
(a)

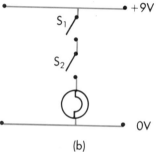
(b)

In circuit (a), there is a more complicated system with two switches, labelled S_1 and S_2, and one bulb. There is an alternative way of representing it in circuit (b). This leaves out the battery and uses two **voltage lines** (or **rails**).

The truth table for this is shown in table 2:

Switch		Bulb
S_1	S_2	
0	0	0
1	0	0
0	1	0
1	1	1

Table 2

This truth table summarizes all of the possibilities. The bulb will only light if both switches are on.

ACTION!

The apparatus of truth

The truth table shown here was obtained using a circuit made up from the same apparatus as in the circuit above but connected differently.

Devise a circuit which would give the results in the truth table.

Inputs		Output
S_1	S_2	
0	0	0
1	0	1
0	1	0
1	1	1

Table 3

Electronic devices

There are many devices used in electronics, some of which are described below:

Light dependent resistors: LDR
These are 'dimmer' switches controlled by light. In strong light LDR allows current to flow through it. In the dark very little current flows.

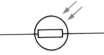

Light-emitting diode: LED
Current flowing through a LED will make it light up providing the current flows in one direction. LEDs use less current and last longer than ordinary bulbs.

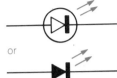

or

Thermistors
These are heat-sensitive resistors.
They conduct electricity better when hot.

Reed switches
These are switches controlled by magnetism. In a glass tube there are two iron strips which are not in contact. When a magnet is brought near, magnetism is induced. The two iron strips are attracted and the switch is closed.

Transistors
These have three connections. The third connection acts as a switch for the other two. If no current flows into the third leg (called the base, b), no current flows between c and e.

Logic gates

These are switches which only allow an output signal in response to certain input situations.

They are called 'gates' because they can be open (logic 1) or closed (logic 0).

Common logic gates are summarized in the table below:

Gate	Truth table			

NOT

Input	Output
0	1
1	0

Sometimes called an inverter.

OR

A	B	Output
0	0	0
0	1	1
1	0	1
1	1	1

Inputs: A, B

Output is high if either input is high

AND

A	B	Output
0	0	0
0	1	0
1	0	0
1	1	1

Inputs: A, B

Output is high if both inputs high

[continued on page 113]

NAND	Inputs		Output	Opposite of AND gate
	A	B		
	0	0	1	
	0	1	1	
	1	0	1	
	1	1	0	

NOR	Inputs		Output	Opposite of OR gate. Output high if neither A nor B is high.
	A	B		
	0	0	1	
	0	1	0	
	1	0	0	
	1	1	0	

ACTION!

Tabling the truth

The diagram is of a circuit with two switches S_1 and S_2.

(a) Complete the truth table:

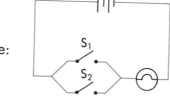

Inputs		Output
S_1	S_2	
0	0	0
0	1	
1	0	
1	1	

(b) What logic gate is represented by this table?

(c) Draw a symbol for this logic gate.

Storing numbers, words and pictures

The microprocessor has enabled us to store, work on and display numbers, words, pictures and sounds. Microprocessors use the simple language of 'on' (1) and 'off' (0); the **binary** code. The microprocessor can store binary code in one of two ways. A permanent store can be made which remains even when the machine is switched off. This is used to store information which is needed to operate the equipment. It is stored in the ROM (read-only memory). Other information can be stored in the RAM (random-access memory) chip. This is lost if the electricity supply is turned off.

A modern calculator works very quickly and can do up to 500 000 sums each second. A large computer can contain up to 250 000 chips.

Numbers are much easier to use than letters. Calculators only have to deal with numbers and so are much smaller than word processors.

The important parts of a calculator are:

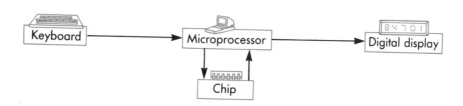

The numbers 0 to 9 can be shown on the digital display by lighting different combinations of seven lights (or liquid crystals). This is shown in below.

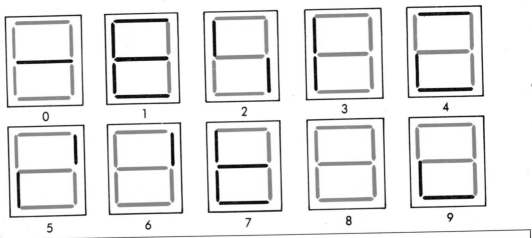

ACTION!

Seeing the light

Complete the following table showing the number of lights on for each numeral:

Numeral	Number of on signals
0	6
1	2
2	
3	
4	
5	
6	
7	
8	
9	

A microprocessor has to be more complicated than a calculator, to store, use and display letters and words. There are 26 letters instead of 10 numbers and you can get small letters as well as capital letters. The principles are the same but more complicated. The input is the keyboard. The microprocessor converts all of the information into binary code. The output is displayed on a visual display unit (vdu) or transmitted to a printer.

Pictures and sounds are more complicated still but once more they have to be converted into binary code.

Control circuits

Look at the circuit diagram below which could control street lights. The bulb (representing the street light) is turned on when the light dims. The NOT switch activates the transistor when the input is turned off.

Light sensors are fitted on street lights, and on automatic lights in factories, etc.

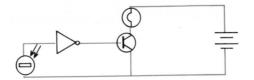

The circuits on page 115 show some simple control devices using two electronic switches. The circuits are controlled by different environmental conditions:

(a) light; (b) heat; and (c) moisture.

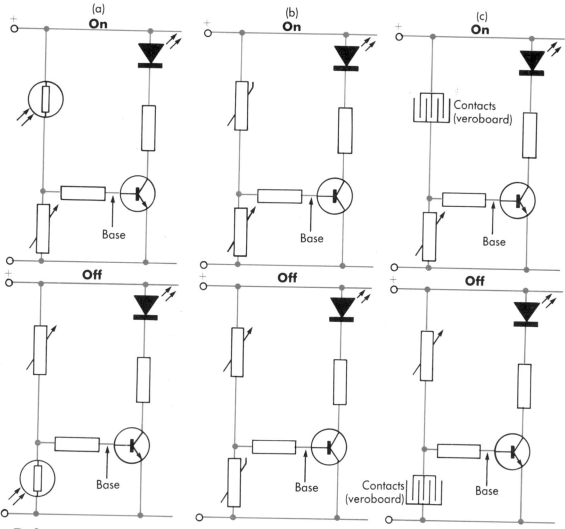

Relays

Relays are electromagnetic switches. Look at the arrangement below of an electromagnetic relay.

The current flowing in the coil causes the soft iron to pivot.

This pushes the two contacts together.

These two contacts are in another circuit and the second circuit becomes switched on.

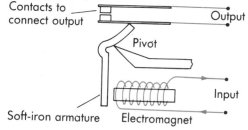

The circuit below shows how a small current in the left-hand circuit can control a much larger current in the right-hand circuit. An example of a relay is in an electronic car ignition system.

When the ignition key is turned on, a small current flows.

The small current flowing operates the relay and switches on the main ignition circuit.

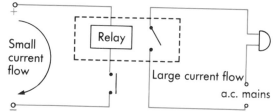

MICROELECTRONICS

There follow on page 116 examples of microelectronic devices used in everyday life:

1 Fire detection A smoke alarm works by sensing smoke inside a special chamber. If smoke is detected the alarm goes off. Many household fires cause death because people are not warned in time to escape.

2 Health There are sensors in hospital equipment that can check continually the condition of a patient.

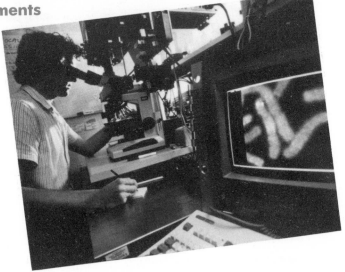

For example, the rate of heartbeat or breathing can be monitored.

A baby in an incubator can be looked after without a nurse having to be present all the time.

3 Microwave oven The control panel on the front of a microwave enables the user to set the time for cooking.

4 Car-engine management Modern car engines are often controlled by microelectronic systems. These enable the engine to be run efficiently.

5 Control of experiments Computers can be used to control experiments.

UNIT 13
ENERGY

You should already know that food provides us with the energy to be active. You should also be familiar with toys having simple mechanisms. A machine needs a source of energy for it to work. You should also be aware that temperature is used to measure how hot or cold things are.

ASPECTS OF ENERGY

Energy makes things happen. It can exist in different forms and can be converted from one form to another. It can be stored in fuels and other chemicals. Energy is measured in units called joules (J), named after the famous scientist James Prescott Joule.

Did you know?

James Prescott Joule (1818–1889)

James Prescott Joule was a Manchester brewer. When he was 19 he built an electric motor and measured the input and output of the motor. Joule was trying to find an alternative to the steam engine. However, the motor was not practical as it took one kilogram of zinc to produce the same work as 200 g of coal.

Joule then built a dynamo and discovered that the electric current produced heat when he turned the winding handle. In order to measure the heat produced he covered the dynamo in water and measured the change in temperature.

From this experiment he set about the task of converting work into heat without first turning it into electricity. He worked for 40 years on this project.

Joule was a very enthusiastic scientist. Even on his honeymoon he tried to measure the temperature at the top and bottom of a waterfall to find out if the water was the same temperature.

(a) Explain why Joule thought that the water at the bottom of the waterfall would be different in temperature from the water at the top.

(b) Would you expect the water to be hotter or colder at the bottom of the waterfall

---

ENERGY IN DIFFERENT FORMS

Energy can exist in seven different forms.

These are summarized opposite.

You can remember them if you remember the name.

L. N. CHEMS: light; nuclear; chemical; heat; electrical; mechanical and sound.

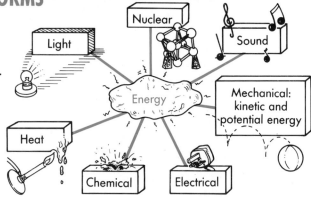

Light energy

Light is a form of energy. Light makes plants grow and powers solar cells. Light causes objects to warm up. Apart from visible light, there are other forms of **electromagnetic radiation**. These include X-rays, ultraviolet (u.v.) and infra-red (i.r.) radiation.

Nuclear energy

Elements such as uranium are radioactive. This means that the nuclei of the atoms split up and emit very large amounts of energy, much of it in the form of heat. In a nuclear power station, this energy is used to produce steam to drive turbines.

Chemical energy

Chemicals can be regarded as **stores** of chemical energy. This energy can be released when the chemical reacts. For example, coal, oil and gas are examples of chemicals which burn and produce heat energy.

A battery for your torch contains chemicals. When the chemicals are used up the battery stops working.

Heat energy

Heat is a form of energy. When ice is heated it melts. On further heating the water boils. When a piece of metal is heated it expands. If it is heated more strongly, it glows and light is emitted.

All matter contains energy and when chemical changes occur there is usually an energy change.

Electrical energy

Electrical energy is a very convenient form of energy because it can be easily converted into other forms. Many appliances are powered by electricity.

Unfortunately, electricity cannot be stored. However, it is possible to store energy in chemicals ready for use when required.

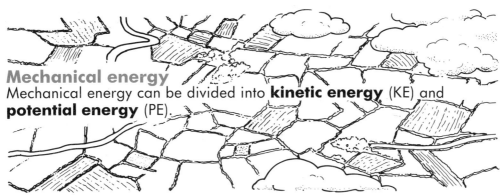

Mechanical energy

Mechanical energy can be divided into **kinetic energy** (KE) and **potential energy** (PE).

The stretched elastic in a catapult contains stored energy. This stored energy is called potential energy. When the elastic is released, the stored energy is released.

An object resting on the ground possesses no energy. When the object is held in the air it has potential energy, (i.e. the potential to be pulled down by gravity).

When the object is dropped, the potential energy is released.

The photograph shows a large waterfall.

The water at the top of the waterfall possesses a large amount of potential energy.

When it falls the potential energy is converted into kinetic energy.

This can be used to generate electricity.

Any object which is moving has kinetic energy. The kinetic energy possessed by an object depends upon the mass of the object and the velocity:

kinetic energy (in J) = ½ × mass (in kg) × velocity² (in m/s)
$$KE = \tfrac{1}{2}mv^2$$

For example, calculate the kinetic energy possessed by an object weighing 100 kg and with a velocity of 20 m/s.

$KE = \tfrac{1}{2} \times 100 \times 20 \times 20$
    $= 20\,000$ J or 20 kJ

A joule is a very small amount of energy. When you walk upstairs you use about 1000 J. Often amounts of energy are given in kilojoules where 1 kJ = 1000 J. Very large amounts of energy may be given in megajoules (MJ) where 1 MJ = 1 000 000 J.

Sound energy

Sound is a form of energy. Sound energy is passed through any material, not just through air. However, unlike light energy, sound energy cannot pass through an empty space (vacuum).

119

ENERGY CHANGES

There are many examples where energy changes from one form to another. The following examples illustrate some of these changes:

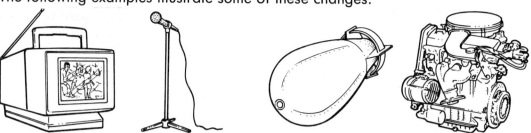

1 Television A television converts electrical energy into light energy, sound energy and heat energy

2 Microphone A microphone converts sound into electrical energy

3 Atomic bomb An atomic bomb converts nuclear energy into heat energy, light energy and sound energy

4 Car engine A car engine converts chemical energy (in the fuel) into mechanical energy, heat energy and sound energy

In many cases, more than one form of energy is produced. Often one form of energy is more important.

Energy change information can be shown by energy arrows as shown.

Chemical energy in petrol → Mechanical energy (kinetic), Heat energy, Sound energy

ACTION!

Energy changes

A microphone converts **sound energy** into **electrical energy**.

For each of the following, state the energy change which takes place:

(a) A bicycle dynamo converts _____ into _____

(b) The brake blocks on a bicycle when the brakes are applied convert _____ into _____

(c) A human body converts _____ into _____

(d) A light bulb converts _____ into _____

(e) A piece of wood when it burns converts _____ into _____

The conservation of energy

The work of Joule around 1850 was revolutionary. Other scientists found it difficult to accept Joule's ideas. He showed that when energy is converted from one form to another, none is lost and none is gained. This idea is summarized by the Principle of Conservation of Energy:

Energy cannot be created or destroyed. It can only be changed from one form to another.

However, some energy conservations appear to lose energy. For example, not all of the chemical energy in the petrol of a car is converted into mechanical energy. We have seen already that some energy in the petrol is converted into forms which do not help to move the car. However, the sum of all the different forms of energy produced must equal the amount of energy stored in the fuel.

All the energy in petrol is not converted into one form (i.e. mechanical energy).

It is spread out and, in this spread out form, it is less useful.

Can you explain into which types of energy it is converted?

Efficiency is a measure of how much energy is transformed from one form to another.

For example: when 1000 kJ of energy is converted into another form, 200 kJ of energy is produced so that:

efficiency = 200/1000 = 0.2

Sometimes this is multiplied to give the answer as a percentage:

efficiency = 0.2 × 100 = 20%

ACTION!

Power crazy

The diagram shows the energy changes in a power station.

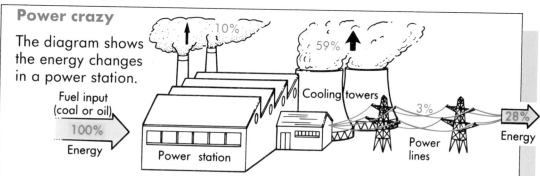

(a) What percentage of the energy of the fuel is:

   (i) lost through the chimney of the power station?
   (ii) lost through the cooling towers?
   (iii) lost in power lines during transmission?

(b) What is the efficiency of this process for converting chemical energy into electricity?

(c) Why is electricity an expensive form of energy?

ENERGY AND WORK

Work is energy being used up. The food you eat is the fuel you need to do everything you have to do. You know that you use approximately 1000 J every time you go upstairs. If you go on a long run, you will use up more energy and do more work.

Did you know?

Marathon runners store up energy by eating pasta, on the eve of the race, that is rich in carbohydrate.

Work can be calculated from the force (in newtons) and distance moved (in metres).

work done (in J) = force (in N) × distance moved (in m)

For example a box requires a force of 50 N to move it.

Find the work done in moving the box 10 m.

work done = 50 × 10
= 500 J

THE TRANSFER OF HEAT ENERGY

There are three ways in which heat can be transferred: convection; conduction and radiation.

Convection Convection can take place in a fluid material, (i.e. a liquid or gas). The heat energy is transferred by movement of the liquid or gas. The movement of the fluid carries heat from place to place. This movement of the fluid is called a **convection current**.

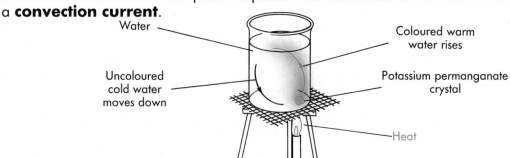

Look above to see a simple experiment that shows convection in a liquid.
A crystal of potassium manganate (VIII), or potassium permanganate, is added to the water. On heating the beaker below the crystal as shown in the diagram, convection currents are set up. The coloured warmer water expands and becomes less dense. This rises and the denser cold water moves in to replace it.

Fires or heaters in a large room can set up similar convection currents.
The heated air rises and cold air falls to replace it. As a result, your feet can be very cold, even though the room is well-heated.

Convection currents are also responsible for sea breezes.

During the day, the land heats up more quickly than the sea. The hot air rises above the land and cool air comes in from the sea to replace it. The result is a breeze blowing onshore from the sea.

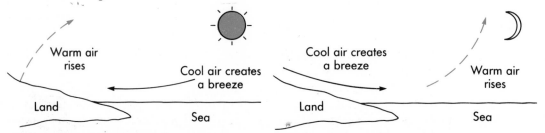

At night the situation is reversed. Now the land cools more quickly than the sea. Convection currents rise from the sea and cool air from the land blows offshore to replace it. The effect is a breeze from land to sea.

Conduction Conduction is the transfer of heat energy through a material. The most important examples of conduction occur in solids, where convection is impossible. Conduction can take place in fluids however.

The heat energy is passed from one particle to the next. The particles do not have to move from one place to another in order to pass on the energy. This is the big difference between conduction and convection.

A substance which passes on heat energy well is called a **good conductor**. Metals are very good conductors of heat. A substance which is a very poor conductor of heat is called an **insulator**.

Look at the experiment to find out which of five metals is the best conductor of heat.

One end of each metal rod is coated with wax.

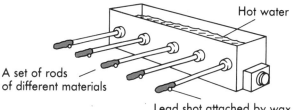

A set of rods of different materials

Hot water

Lead shot attached by wax

The other end of each rod is heated in the same hot water bath. (This ensures the metal rods are heated the same.) The best conductor of heat will be the metal where the wax starts to drip first.

Radiation The energy coming to the Earth from the Sun cannot come by convection or conduction. This is because convection and conduction cannot pass through the empty space between the Sun and the Earth. Radiation is energy transferred directly from the source to the object being heated.

There are many different kinds of radiation but, when we are considering heating effects, the radiation is called **infra-red radiation**. The energy is given out by a **source** such as the Sun, in this case. The Sun radiates energy because it is very hot. This energy is felt directly on the Earth.

In a similar way, if you hold your hand near an electric fire, your hand feels warm as the fire radiates heat in all directions.

Radiation is direct and travels in straight lines. It can travel through empty space.

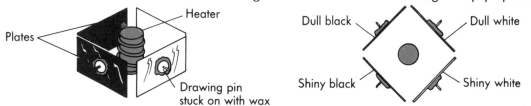

Plates

Heater

Drawing pin stuck on with wax

Dull black

Dull white

Shiny black

Shiny white

Different surfaces absorb radiation to different extents. Look at the above experiment to compare the amount of radiation absorbed by four different surfaces. A drawing-pin is fixed to the back of each sheet with wax. The drawing-pin will drop first from the sheet which absorbs heat best.

The results would show that dark, dull colours are the best absorbers of heat energy whereas bright, shiny colours are the worst absorbers.

ACTION!

The heat of the moment

For each of the diagrams below, state whether the method of heat transfer is convection, conduction or radiation.

A
Copper rod
Paper does not burn

B
Cardboard
Burning paper produces smoke
Candle
The arrows show the direction of the smoke

C
Paper spiral

D
Metal disc stuck on with wax
Shiny surface
Black Surface

E
Rods of different materials
Lead shot attached by wax
Heat

123

Waste not, want not: heat loss at home

In winter, about one third of the energy in Great Britain is used for home heating. Much of this energy is lost due to inadequate insulation. The diagram summarizes the heat loss in a typical home.

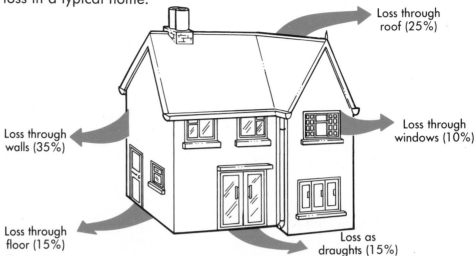

Loss through roof (25%)

Loss through windows (10%)

Loss through walls (35%)

Loss through floor (15%)

Loss as draughts (15%)

There is much that can be done to reduce this heat loss from homes. This will reduce the fuel costs.

1 Loft insulation Heat loss through the roof can be reduced by insulation with a mineral wool. This is laid between the joists in the loft space. Air is trapped between the fibres of the mineral wool and this makes the material a good insulator. This layer of insulation prevents heat loss by conduction. For good insulation, a thickness of 15 cm is recommended.

2 Sealing draughts Draught excluders on doors prevent currents of cold air entering a house. It is, however, essential that there is adequate ventilation for open fires and gas fires if they are to burn safely without producing poisonous fumes.

3 Double glazing You will notice in the diagram that only a relatively small amount of heat is lost through the windows. This loss of energy can be reduced by fitting double glazing. Such windows have two panes of glass with air trapped between them.

The trapped air acts as a good insulator and prevents heat escaping. Double glazing can be expensive but it does reduce condensation and soundproofs the house.

Two panes of glass

20 mm gap

4 Cavity wall insulation In most houses, walls have two vertical parts with a gap or cavity between them. The air in this cavity can move about and convection currents can be set up. If the cavity is filled with polystyrene beads or foam, which are both good insulators, these currents are prevented. Much less heat is then lost through the walls.

124

5 Carpets and underlay Heat loss through the floor can be reduced by fitting thick carpets and underlay. Both of these trap air and act as good insulators. This prevents conduction through the floor.

Can you think of any other ways of saving heat?

Did you know?

Comparing fuels

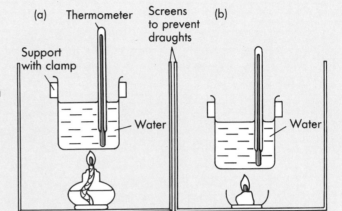

Look at this simple experiment to compare the energy produced when two fuels are burnt.

The first fuel is methylated spirit, contained in a small lamp.

The second fuel is a small piece of fire-lighter.

In each case, the amount of fuel used can be obtained by weighing before and after. The table below gives the results for the two fuels.

	Methylated spirit	Fire-lighter
Mass of water	100 g	100 g
Mass of fuel used	0.5g	1.0 g
Temperature before	20°C	20°C
Temperature after	43°	35°C
Cost of fuel in p per 100 g	5	1

Table

In selecting the best fuel for a particular purpose, price is an important consideration. However, there are other things to consider. These include:

1 How readily available is the fuel?

2 How cleanly does it burn? Does it produce unpleasant fumes?

3 Is it a solid, liquid or gas? How is it delivered and stored?

4 How easily does it catch alight?

ACTION!

Fuel for thought

(a) Use information in the table above to find which fuel produces the most energy per gram burned.

(b) Which fuel is most economical? (i.e. most energy per penny)

Fuelling the energy debate

The diagram shows the quantity of energy produced when 1 kg of different fuels are burnt.

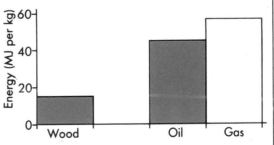

(a) Complete the diagram by adding the information for coal. Coal produces 29 MJ per kg.

(b) Which fuel produces the largest amount of energy per kg?

FOSSIL FUELS

Fossil fuels include coal, petroleum and natural gas. Coal was produced by the action of heat and pressure on trees and plants over millions of years. The trees and plants obtained their energy from the Sun. Petroleum and natural gas were produced by the action of heat and pressure on tiny sea creatures over millions of years.

Fossil fuels took millions of years to produce. Unfortunately, they are being used up rapidly. The amounts of these fossil fuels in the Earth are limited. Therefore, it is important to find alternative energy sources and to use existing energy sources carefully.

Wood is a **renewable fuel**. Every year new supplies can be grown and with sensible use it should never run out. However, due to necessity and even greed in some parts of the world, more timber is being felled than is being grown. In Brazil, ethanol (alcohol) is being produced by fermentation of sugar. Every year new sugar can be grown and fresh stocks of ethanol can be produced. Ethanol is another renewable fuel.

When fossil fuels burn they produce carbon dioxide. The increased use of fossil fuels is increasing the percentage of carbon dioxide in the atmosphere. This could be very serious in the future.

Radiation passes through the Earth's atmosphere and heats up the planet. As the Earth heats up, it starts to radiate energy of a shorter wavelength. This radiation cannot pass through the carbon dioxide in our atmosphere and escape. Thus, the temperature of the Earth rises which can alter the climate of the Earth. It is predicted that the ice caps at the pole could melt partially causing sea levels to rise and land to be submerged. In Great Britain, the summers could become hotter and dryer and the winters colder and wetter. These changes could affect the way people live. These changes are usually called the **greenhouse effect**. Look out for articles in newspapers and magazines about this important issue!

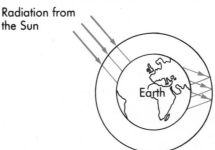

Radiation from the Sun

Radiation from Earth of lower wavelengths do not pass through the carbon dioxide in the atmosphere

Earth

ACTION!

Hot under the collar

Read the following article from the *Daily Telegraph* about the possible effects of global warming.

A touch of the tropics is coming our way

THIS MONTH a group of Government-sponsored scientists began an unpublicised and sensitive project. Their brief: to make informed guesses about what would happen to the economy, agriculture and ecology of Britain if, by the year 2050, the global climate has been drastically altered by the heating process known as the greenhouse effect.

Assume, the scientists were told, that the global average temperature rises by 1·5 to 4·5°C in 60 years — a lightning shift in the life of the Earth since only 5°C separate us from the last Ice Age. Assume, too, a rise in sea level of 0·2 to 1·4 m and variations in rainfall of 50 cms a year.

What would happen? Would Britain be able to grow the same crops, the same trees? Could we use the same water supplies? Would the country sustain the same wildlife? We may assume that the answer is probably not.

For some years environmental Cassandras have been ringing alarm bells by predicting, for example, the gradual melting of the polar icecaps — but the evidence of a speedy onset of global warming has been less than firm.

Britain's £250,000 "desk-top" project, which is to report in six months, is one of many indications that the greenhouse effect has begun to trouble Western governments. "The effect is real, there is no doubt of that," a senior Government scientist told me last week.

It seems that having looked into the projections of, among others, the University of East Anglia's climate research unit and the Met Office computer models, the Government has decided to take precautions.

What the "desk top" studies now commissioned are likely to predict is, as expected that whenever the warming happens, many of our sea defences would need to be rebuilt much higher. Water supplies might become salinated. New planning regulations might be needed to prevent houses flooding.

Unpredicted until recently

ENVIRONMENT
Charles Clover

are some of the possible climatic changes. In the next century central England could acquire a climate rather like Bordeaux. Red wine-growing would become possible. Forests would be the least likely to adapt, bringing devastation to cold-weather loving conifer plantations. Many wildlife species would be forced to move northwards to find a more temperate climate. Many plant species, from crocuses to apple trees, would die out.

Some evidence suggests higher average temperatures might actually mean more extremes: harsher frosts and hotter summers, more storms.

Britain, however, looks to be lucky. For low-lying countries such as Bangladesh rising sea levels could mean worse floods than those which already regularly devastate the Ganges basin.

Changed rainfall patterns could mean new deserts in Africa and dustbowls in the Mid-Western states of America, the Soviet Union, and perhaps the thickly populated Nile delta.

The greenhouse effect is caused by the build-up in the Earth's atmosphere of gases — principally carbon dioxide from the burning of wood and fossil fuels. These gases are likely to form a layer of "double glazing" through which radiation from the sun will be able to enter, but reflected warmth from the earth will not be able to escape — so the Earth's atmosphere will heat up. Other "greenhouse" gases are the ozone-depleting chemicals CFCs and methane.

Few scientists believe there is yet definite evidence that it has begun to happen. But, for one reason or another, global average temperatures have risen by 0·5°C since the middle of the 19th century. Four of the hottest years on record occur in the 1980s, the hottest of all being 1987.

Huge assumptions have to be made about how fast we will reach the worrying doubling of carbon dioxide, since such questions depend on the energy needs of developing countries.

Research needs to be done into how any carbon dioxide increase would heat the climate, taking into account the buffering effect of oceans absorbing carbon dioxide. But our best guess brings us back to the above brief — given to the biologists and scientists doing the desk-top studies.

Department of the Environment scientists believe we are still five years away from an understanding of temperature rise, 15 years away from predicting climatic effects in, say, Britain; 30 years away from having the greenhouse effect "tied down" in accurate computer models.

That is a long time considering the carbon dioxide the world's energy industries are likely to generate in that time — and the greenhouse effect is not reversible, at least for thousands of years.

Some optimism now exists that a global agreement might perhaps be reached to limit carbon dioxide levels in time — as a result of last year's first global anti-pollution agreement to reduce CFCs. An international meeting to discuss the greenhouse effect takes place in Montreal this July.

Nuclear power, which creates no CO_2 but is expensive, together with energy conservation, is likely to be discussed. Britain is inclined to criticise certain Western countries who want action on the greenhouse effect — but still subsidise electricity use.

The biggest priority of all is research into climate modelling and, particularly, the oceans. Such research in Britain has been under strict spending limits, according to the perenially empty begging bowls of the Natural Environment Research Council. When the results of the "desk-top" reports flood in, that may change.

[continued on next page]

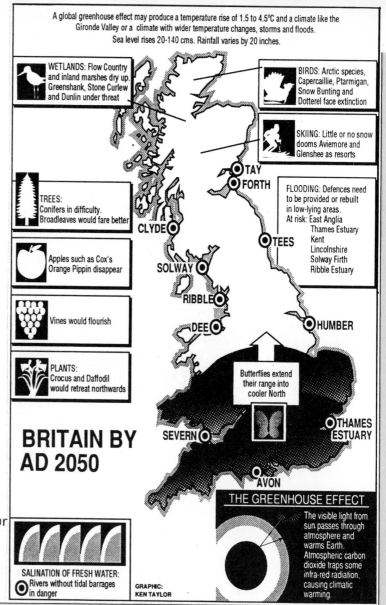

A global greenhouse effect may produce a temperature rise of 1.5 to 4.5°C and a climate like the Gironde Valley or a climate with wider temperature changes, storms and floods. Sea level rises 20-140 cms. Rainfall varies by 20 inches.

WETLANDS: Flow Country and inland marshes dry up. Greenshank, Stone Curlew and Dunlin under threat

BIRDS: Arctic species, Capercaillie, Ptarmigan, Snow Bunting and Dotterel face extinction

SKIING: Little or no snow dooms Aviemore and Glenshee as resorts

TREES: Conifers in difficulty. Broadleaves would fare better

Apples such as Cox's Orange Pippin disappear

Vines would flourish

PLANTS: Crocus and Daffodil would retreat northwards

FLOODING: Defences need to be provided or rebuilt in low-lying areas.
At risk: East Anglia
Thames Estuary
Kent
Lincolnshire
Solway Firth
Ribble Estuary

Butterflies extend their range into cooler North

BRITAIN BY AD 2050

SALINATION OF FRESH WATER: Rivers without tidal barrages in danger

GRAPHIC: KEN TAYLOR

THE GREENHOUSE EFFECT
The visible light from sun passes through atmosphere and warms Earth. Atmospheric carbon dioxide traps some infra-red radiation, causing climatic warming.

TAY, FORTH, CLYDE, TEES, SOLWAY, RIBBLE, DEE, HUMBER, SEVERN, THAMES ESTUARY, AVON

Use the information in this article and any other information you have to write a letter to your Member of Parliament stating your concerns about the effects of global warming and any suggestions you have to reduce the effects.

RENEWABLE FORMS OF ENERGY

As deposits of fossil fuels start to run out and become more expensive, alternative sources of energy must be found.

Solar energy More energy reaches the Earth's surface from the Sun in an hour than is used by the world in a whole year. Even in Britain, energy from the Sun is 80 times our present energy demand.

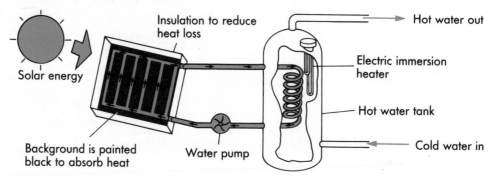

Solar energy

Background is painted black to absorb heat

Water pump

Insulation to reduce heat loss

Hot water out

Electric immersion heater

Hot water tank

Cold water in

At present solar energy is not used very well. Only a few houses, hotels and factories have solar panels fitted on the roof. These panels are painted dull black to make them better absorbers of radiant energy.
The solar-heated water in the panels is then circulated and used to pre-heat the water in the hot-water tank. Less fuel is needed to heat the water.

128

Solar cells can be used to convert radiation directly into electricity. Calculators and watches can be powered in this way. In developing countries, this kind of technology can be used to produce electricity to pump water or light lamps.

Did you know?

The Nixon Memorial Hospital in Sierra Leone has solar panels to generate electricity for lights in wards, pumping water and sterilizing instruments. The electricity is stored in lead-acid batteries. Due to the high cost of diesel oil, the generator can only be used for three hours each evening and for emergency operations.

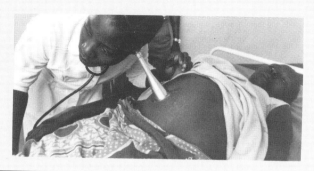

A more complicated method involves putting the solar cells into orbit around the Earth. They can be more effective there. The radiation is then converted into microwave energy. This is transmitted back to a ground station where it is converted into electricity.

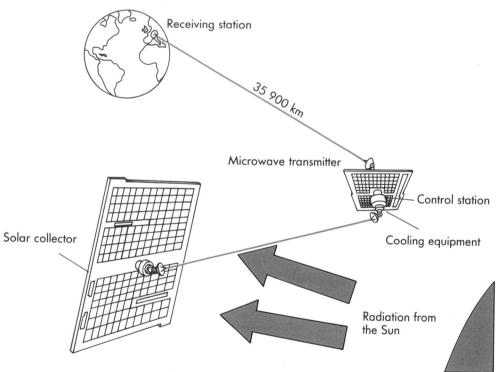

Receiving station

35 900 km

Microwave transmitter

Control station

Solar collector

Cooling equipment

Radiation from the Sun

Wind energy Wind energy has been used in the past for windmills.

With new technology it is now possible to collect energy from the wind more effectively and convert it into electricity.

Look at the modern wind turbine.

A whole series of these could produce up to 20% of the electricity we need in Britain.

Hydroelectric energy Study how energy changes in a hydroelectric power station. Electricity is generated by water falling from the reservoir. Its potential energy is converted into kinetic energy. The falling water turns a turbine which generates electricity.

Clouds

Rain

Reservoir Dam Electricity supply

Water pipeline Turbine house

ACTION!

Hydro-generation

The map shows where the hydroelectric power stations are located in England, Wales and Scotland.

(a) In which areas are the hydroelectric power stations located?

(b) Why are these areas suitable for hydroelectric power stations?

(c) What is the rainfall like in these areas?

(d) Would you expect hydroelectric power stations to be built in the Netherlands?

Give a reason to explain your answer.

Tidal energy In the estuary of the River Rance in France, there is a power station which generates electricity using tidal power.

The working of the power station is shown opposite.

A similar dam across the Severn Estuary could produce 20% of our electricity.

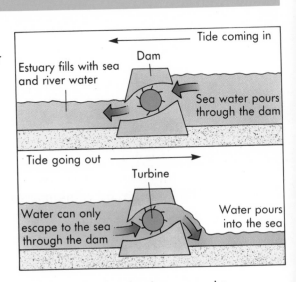

Tide coming in

Dam

Estuary fills with sea and river water

Sea water pours through the dam

Tide going out

Turbine

Water can only escape to the sea through the dam

Water pours into the sea

Wave energy The movement of ocean waves can also be harnessed to produce energy. Experiments have been carried out using large rafts which move up and down with the waves. This rocking movement can be converted into electricity.

130

Geothermal energy This is not a renewable source of energy. It relies on the fact that the rocks beneath the Earth's surface are much hotter than those at the surface. Countries such as France, Hungary, Japan and New Zealand rely on geothermal energy for some of their energy needs. Water is piped below ground and then pumped back up to the surface. The heated water can be used for central heating and various other purposes.

ACTION!

The Swedish solution

Read the following passage about alternative fuels in Sweden and use it to answer the questions which follow.

Sweden has no deposits of coal or oil. Most of its electricity is produced from nuclear power. However, the accident at Chernobyl has made extending nuclear power unpopular with people.

One of Sweden's greatest resources is the conifer forests which are being extended to cope with demands for paper and softwood.

Studies have been made to find trees which grow quicker than conifer trees which can be cropped quickly for fuel. Alder, poplar and birch are three that have been tried. Finally, it has been decided that the willow tree is most suitable. Willow trees: grow quickly; contain very little nitrogen and sulphur; and produce ash on burning which is rich in potassium.

Willow trees are being planted about one metre apart in plantations. After four years the trees are cut back leaving just stumps. The stumps grow again and a fresh crop can be taken every four years.

All of the wood cut is turned into willow chips which can be easily burned.

Just like burning fossil fuel, such as coal and oil, burning wood produces carbon dioxide. Carbon dioxide contributes to the greenhouse effect. However, it can be argued that the carbon dioxide produced by burning willow chips is only the same as the amount of carbon dioxide taken in by the tree when it grows.

(a) Why does Sweden not use conifer forests as a source of fuel?

(b) What are the three advantages of willow trees as a source of wood for fuel?

(c) Why should a tree which is low in nitrogen and sulphur be used as fuel?

(d) What can the ash produced by burning willow chips be used for?

(e) What is the process taking place in the tree when it absorbs carbon dioxide?

GB sources

The pie chart opposite shows the energy sources used in Great Britain in 1988.

(a) Which of these sources of energy:

  (i) is most widely used
  (ii) is a renewable source of energy
  (iii) does not produce any pollution
  (iv) is a solid fossil fuel

(b) What changes in the pie chart might be expected if the price of oil were to increase greatly?

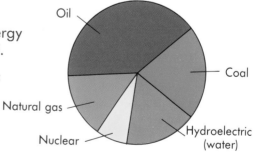

How might greater use of alternative energy sources change your daily life?
What might happen to countries which have fossil fuels to sell to other countries?

PROJECT

Travelling in cities Throughout the world many cities are being clogged up with cars, lorries and buses. These vehicles use up petrol and diesel fuel. They also use up oxygen when the fuels burn. They are inefficient converters of chemical energy into mechanical energy. They also produce pollutant gases such as carbon monoxide. In Tokyo, Japanese policemen on traffic duty have to wear breathing masks to avoid being poisoned.

Attempts are being made to reduce problems caused by the internal combustion engine. Some of these include: fitting catalytic converters to exhaust pipes; using battery powered vehicles; introducing light railways that run on existing roads; building overhead or underground railways; introducing bus lanes and cycle lanes; charging people to use the roads in cities; and allowing cars into cities on alternate days only, (e.g. Athens)

(a) Find out what you can about these and other ways of improving transport in cities. Write down the advantages and disadvantages of each of these methods?

(b) Why do people prefer to use their own cars in cities rather than public transport?

(c) Apart from pollution, why do cars cause particular problems in cities?

UNIT 14
SOUND AND MUSIC

You should already know that sounds can be made in a variety of ways and that, in an orchestra, different instruments produce sounds in different ways.
A sound is heard when it reaches your ears.

SOUNDS AROUND

Think of a world without sound. You would not be able to hear: an orchestra playing; the gentle rustling of leaves in a tree; the roaring of traffic along a busy road; or a jet airliner taking off. Some sounds are pleasant while others are unpleasant and possibly dangerous. Make a list of sounds which are members of these two sets.

All sound is made by vibration. This is the movement caused when an object travels backwards and forwards very quickly. When an object vibrates, it disturbs the air around it in all directions. When you switch on a radio, its loudspeaker sends ripples of sound through the air in all directions. These ripples are called **waves**.

ACTION!

> ### How quickly does sound travel through the air?
>
> A friend standing one kilometre from you fires a starting pistol. You see the smoke from the starting pistol and three seconds later you hear the bang. Light travels instantaneously from the gun to you but sound takes some time. Calculate the speed of sound.
>
> **Taking soundings**
>
> (a) When a ship fires its gun, the crew hear an echo five seconds later from a cliff face opposite. If the speed of sound is 300 m/s, what is the distance of the ship from the cliff?
>
> (b) In a cathedral, a person sitting 165 m from the organ and 5 m from the nearest loudspeaker notices a time delay of 0.5 s between the music transmitted by the amplification system and that transmitted from the organ through the air. Calculate the speed of the sound in the air in the cathedral. State any assumptions you have made.

Sound waves move quickly through air (about 340 metres each second at 20°C).

Sound also travels through other substances but its speed is different in different materials. It depends upon the density of the substance through which it travels.

Sound travels about 0.5 metres per second faster for each one degree rise in temperature.

ACTION!

Speed of sound through different materials

The table shows the speed of sound through different materials at 20°C:

Material	Speed (m/s)
air	340
steel	6000
softwood	
water	1461
brick	3600
aluminium	5100

(a) Complete the bar chart of the speed of sound in different materials.

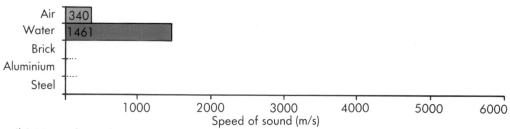

(b) How does the speed of sound vary with the density of a material?

(c) What would you expect the speed of sound to be through polystyrene? Explain your answer.

(d) Which of the following could be the speed of sound through softwood? 200, 1200, 3200, 5200

(e) What do you think is happening in the cartoon?

He placed a bell inside a glass container and set the bell ringing.

He then used a pump to suck all of the air out of the container.

The bell could still be seen vibrating but no sound could be heard. The he let the air slowly back into the container. The sound of the bell ringing came back. This experiment demonstrated clearly that air was needed in the container for the sound to pass from the bell to the ear.

Did you know?

Sound cannot pass through a vacuum. Due to this, an explosion in outer space could not be heard on Earth.

The apparatus below is similar to that used by Hawsby. For the sound to be heard, it had to travel through air, then glass, and finally air again before it reached the ear. When the bell vibrates it makes the air next to the bell vibrate.

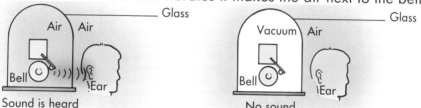

The movement of this air produces a wave in which the particles move backwards and forwards. This type of wave is called a **longitudinal wave**.

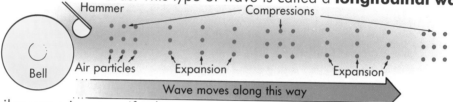

A similar wave is set up if a long spring is pushed quickly at one end. The alternate places of compression and expansion travel along the spring to the other end.

The wave moves from one end of the spring to the other carried by the particles in the spring.

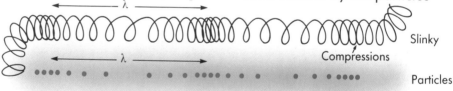

Sound normally travels in all directions. However, sometimes it can be directed by using a megaphone.

Echoes

In a large, empty building you may hear echoes when you shout. Echoes are caused by sound waves bouncing back off the walls. This can cause problems in concert halls. Echoes can be reduced by curtains, carpets and furniture and by using padding on the walls. Why do you suppose that these measures reduce echoes?

Echoes can be useful. SONAR was developed during World War II to detect submarines. Sound waves are bounced back off the submarine so that the submarine can be detected.

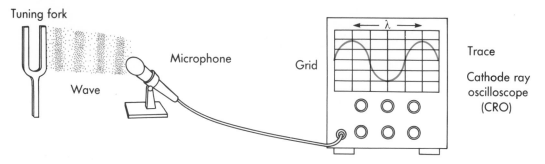

Ultrasound is high energy sound which is produced when a crystal vibrates. When a narrow beam of ultrasound is directed at a person's body, the bones, muscles and organs bounce the sound back in different ways. From the results, a doctor can make decisions about how to treat a patient.

FREQUENCY, PITCH AND VOLUME

Sound waves can be detected by a microphone. The microphone converts the sound waves into electrical signals. These signals are shown on the screen of an oscilloscope.

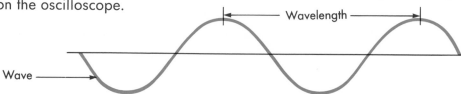

A tuning fork vibrates to produce a single musical note which is pure and gives a wave on the oscilloscope.

The distance between the same point on adjacent waves is called the **wavelength**. In one wavelength the wave goes through one complete vibration. The number of wavelengths or vibrations per second is called the **frequency**.

The **pitch** of a sound depends upon the frequency.

Few vibrations = low frequency = low pitch
Many vibrations = high frequency = high pitch

One complete wave
Low frequency
Low pitch
Long wavelength

Three complete waves
Higher frequency
Higher pitch
Shorter wavelength

When one complete wave is produced each second, the frequency is one hertz (Hz). The human ear can detect frequencies between 20 and 2000 Hz. As people get older, they may lose the ability to detect all frequencies. Dogs can detect much higher frequencies than humans. They can detect whistles which we cannot hear. How could you investigate this fact?

The volume of a sound is not the same as its wavelength, pitch or frequency. Volume means intensity. Look at the waves produced by two sounds of different volume. In the two sounds, the pitch and wavelength are the same. Notice that the height (or depth) of the wave is different.
This is called the **amplitude**.

The louder the sound, the more energy it has and the greater is the amplitude.

a = amplitude

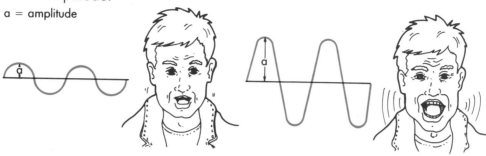

MUSICAL INSTRUMENTS

Many musical instruments make sounds by having vibrating strings. These include guitars, violins, harps and pianos. Sometimes the sound that is heard is **amplified** by a wooden sound box. This means that the sound is made louder by increasing the amplitude. There are three things that affect the sound produced by a vibrating string:

1 The length of the string Long strings vibrate at low frequencies and short strings at higher frequencies. A piano has strings of different lengths to produce a range of frequencies.

2 The tension of the string The more stretched a string is the higher the frequency. When musicians 'tune up' they alter the tension of the strings so as to get the pitch exactly right.

3 The thickness of the string Thick strings give lower frequencies than thin strings because they vibrate more slowly.

Compare the sounds of a string and one of half its length. What do you notice?

Other musical instruments make sounds by allowing air to vibrate in tubes of different lengths. These include recorders, trumpets and organs. When air vibrates in a short, thin tube a very high note is produced. A deep note is produced in a long thick tube. A church organ has many pipes of different sizes to produce a range of different notes. In a recorder, the effective length of the column of air is altered by changing the positions of the fingers. So a range of notes can be produced by the one tube.

ACTION!

Getting to the pitch

How does the musician change the pitch of the note being played?

THE HUMAN EAR

The human ear is a very delicate instrument. It controls our sense of balance as well as our hearing. Sound waves are collected by the outer ear.

The waves pass into the middle ear and make the ear drum vibrate. The movement of the ear drum is carried through the air-filled middle ear by the anvil, hammer and stirrup. These are three tiny bones which pass on the vibrations to the oval window at the innermost side of the middle ear. The cochlea is a coiled tube filled with fluid. Vibrations of the oval window produce pressure changes in the fluid in the cochlea. The fluid affects tiny hairs which cause impulses to be sent along the auditory nerve which carries messages to the brain. The brain interprets these messages as sounds.

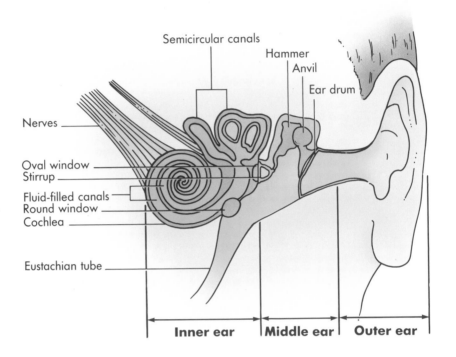

Semicircular canals
Hammer
Anvil
Ear drum
Nerves
Oval window
Stirrup
Fluid-filled canals
Round window
Cochlea
Eustachian tube

Inner ear | **Middle ear** | **Outer ear**

NOISE AND THE ENVIRONMENT

When considering sound, and the effects of sound, people often confuse intensity and loudness. The **intensity** means the amount of energy in the sound waves whereas **loudness** is the apparent strength of the sound at the eardrum which is transmitted to the brain.

The unit of intensity of sound is the **decibel** (dB).

The chart shows the intensities of certain sounds.

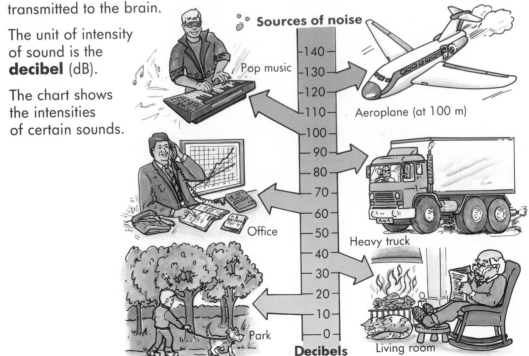

Sources of noise

Pop music
Aeroplane (at 100 m)
Office
Heavy truck
Park
Living room

140 130 120 110 100 90 80 70 60 50 40 30 20 10 0

Decibels

The intensity of sound can be measured using a sound level meter.

The table shows the intensity of the sound produced by different musical instruments at a distance of 3 metres in the open air.

Instrument	Decibels (dB)
clarinet	86
piano	94
trumpet	94
trombone	107
bass drum	113

Not to scale

Traffic produces a great deal of noise. All petrol and diesel engines are fitted with silencers. Americans call them mufflers which is a better word to explain how they work. They reduce the noise rather than remove it. An efficient silencer can reduce the noise level from 160 dB to an acceptable 85 dB.

Cars are soundproofed to prevent engine and road noise entering the body of the car. The engine is often mounted on rubber to reduce the vibrations. Felt is put around the engine compartment.

In cities, motorways are often sunk into cuttings in the ground to deflect the sound upwards and away from nearby buildings.

Television and recording studios are soundproofed to prevent sounds from outside being heard. In addition, windows are double glazed.
The two layers of glass help to reduce the sound level inside the studio.
Walls in the studios are covered with cork, felt or other insulating materials.

Loud noises can damage our hearing especially if we are exposed to it for long periods. Loud noises also speed up pulse and breathing rates.
Recent research also suggests that loud noise can cause heart attacks, ulcers and other conditions.

People working in noisy environments are encouraged to wear ear protectors.

UNIT 15

LIGHT

You should already know that light comes from different sources and can have different colours. For example, Christmas tree lights are different colours. You should realize that light passes through some substances, like glass, but not through others. When light does not pass through an object, a shadow may be formed.

AND THEN THERE WAS LIGHT

An object which gives out light is said to be **luminous**. One which does not give out light is **non-luminous**. A lighted candle, the Sun and fires are examples of luminous objects.

Light from a luminous object travels in straight lines. Sometimes a narrow ray of light passes through an object. The object is then said to be **transparent**. Colourless glass is a transparent material. The particles in the glass do not absorb any of the light and so it passes through the glass unchanged.

Light cannot pass through many materials because the particles in the materials absorb the light. These substances are said to be **opaque**. Examples of opaque materials are wood and copper.

Light is emitted in all directions from a luminous source. The light travels in all directions from the source. If the light hits an opaque object it stops. If the object has a sharp edge there will be a sharp cut-off for the light.

The object will produce a dark **shadow** on a screen.

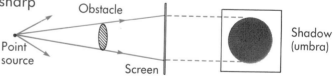

A dark shadow which has the shape of the object is called an **umbra**. If the light source is large, an area of partial shadow appears on the screen.

This is called a **penumbra**.

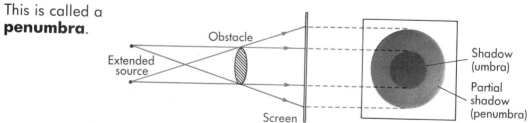

ACTION!

Shadows

You need a light, a mug, a screen and a partially darkened room for this investigation. Place the mug between the light and the screen to make shadows of the mug on the screen.

How would you produce a large shadow on the screen?

Shadows produced by the Moon and the Earth are considered in Unit 16.

SEEING THINGS: REFLECTION

When light hits a rough object which is not transparent, some of the light is absorbed. The rest of the light is **scattered** in all directions.

This is sometimes called **diffuse reflection**.

Some of the scattered light enters our eyes and we see the object. If the surface is smooth and shiny the light is reflected in one direction (e.g. a piece of metal). This is called **regular reflection** or just **reflection**. An ordinary flat mirror is very good at reflecting light rays. It is called a **plane mirror**.

Look at what happens when a ray of light strikes a plane mirror. The light ray which strikes the mirror is called the **incident ray** and the ray which leaves the mirror is called the **reflected ray**.

The line at right angles to the mirror is called the **normal**.

The angle of incidence (i) and the angle of reflection (r) are always equal.

Mirror Normal Back
Front
Reflected ray r i Incident ray
These two angles are **always** equal

The image appears as far behind the mirror as the object is in front of it. The image is reversed or back-to-front. Look at how the image is formed. The ray from 0 hits the mirror and gets reflected into the viewer's eye. But the eye thinks that the light has come from the point I. This is because light is expected to travel in straight lines. So, the image at I is **virtual**. This means it can be seen but it cannot be focused on a screen.

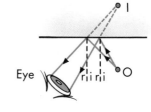

Eye r i r i O

Back to front: lateral inversion

The image produced by a plane mirror is the same way up but the wrong way round. This is called **lateral inversion**.

ACTION!

Reflections in a mirror

A number of letters of the alphabet are **symmetrical**. This means that you can draw a line through the letter in such a way that one side of the line is a reflection of the other.
For example:

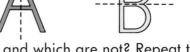

Which capital letters are symmetrical and which are not? Repeat the exercise with small letters as well?

If you place a mirror on a letter H, how can you arrange it so that the image in the letter completes the letter?

Try this for other capital letters too!

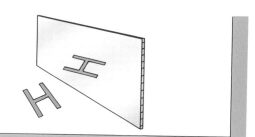

Reflections on curved mirrors

Mirrors which are not flat cause the image to be distorted.

This is used in fairgrounds for amusing effects.

Curved mirrors are frequently used. They can be divided into two groups.

Concave   Convex

Convex mirrors are used: in shops for security; for wing mirrors on cars; or on the top deck of a bus, so that the conductor can see down the steps.

A convex mirror produces an image which is: **virtual; upright;** and **smaller** than the object.

Look at the image formed by a convex mirror.

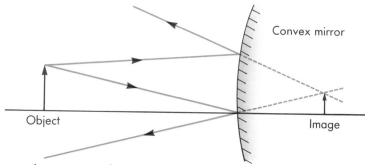

Concave mirrors are also widely used. If an object is held close to a concave mirror the image is: **virtual; upright;** and **larger** than the object.

When a dentist is examining your teeth with a mirror, a concave mirror will magnify them so that any decay can be clearly seen.

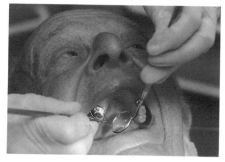

If the object is moved a long way from the concave mirror the image changes. It becomes: **real; upside down;** and **smaller** than the object.

Look at how this image is formed.

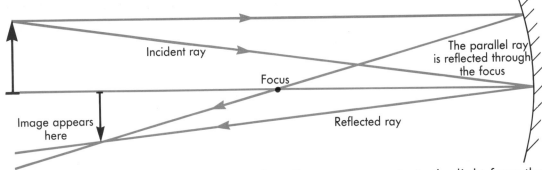

A concave mirror is also used in a car headlamp to concentrate the light from the headlamp bulb.

Think about why a concave or convex mirror is used for each example!

REFRACTION

Light passes at different speeds through different materials. When light passes from one material to another bending occurs. This is called **refraction**.

Light rays bend when they enter a glass block and bend again when they leave the glass block. Light refracts towards the normal as it enters the glass. On leaving the glass it bends away from the normal. If the light strikes the glass block at 90°, the ray passes through the block unchanged.

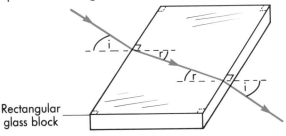

Rectangular glass block

Refraction can distort the view of objects we see underwater. Refraction of light rays leaving the water make objects seem less deep than they really are.

Find out the reason why mirages occur.

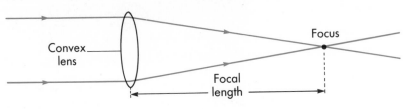

LENSES

Lenses are pieces of plastic, or glass, with one or both faces curved. When light passes through a lens, the light is refracted at both faces. The path of light is altered depending upon the curvature of the lens.

Lenses that are thicker in the middle than at the edges are called **convex** lenses. They will make the light rays converge and so come closer together.

When parallel light rays pass through a convex lens the rays are brought together at a **focus** F. The distance between the centre of the lens and F is called the **focal length**. The focus of a convex lens is a real focus because the light rays from the lens actually pass through the focus.

Lenses that are thinner in the middle than on the edge are called **concave** lenses. A concave lens causes light rays to diverge.

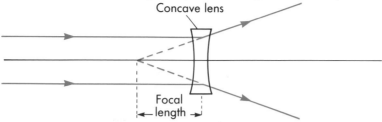

Parallel rays passing through a concave lens seem to have come from a focus F. Again, the distance between the centre of the lens and the focus is called the focal length. However, this time, the focus is virtual as light only appears to come from this point, the rays do not actually pass through it.

Image formed by a convex lens

The nature and position of the image formed with a convex lens depends upon the position of the object relative to the focal length of the lens.
The situation is summarized in the table below.

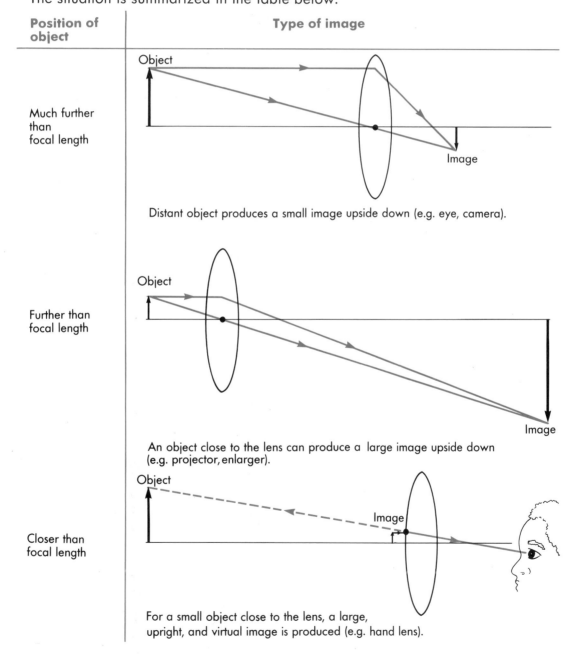

Position of object	Type of image
Much further than focal length	Distant object produces a small image upside down (e.g. eye, camera).
Further than focal length	An object close to the lens can produce a large image upside down (e.g. projector, enlarger).
Closer than focal length	For a small object close to the lens, a large, upright, and virtual image is produced (e.g. hand lens).

Image formed by a concave lens

A concave lens produces a virtual, upright image which is smaller than the object. The **ray diagram** describes how the image is formed.

The light ray from the top of the object passing through the centre of the lens is unchanged in direction.

The light ray at right angles to the lens is bent as shown.

Both rays appear to come from I.

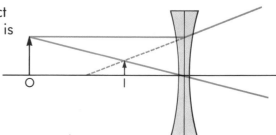

ACTION!

Portraying an image

The diagram shows a pinhole camera. This is the simplest form of camera since it does not use a lens.

The lamp is 10 cm tall and is placed 20 cm from the camera.

The camera is 20 cm long.

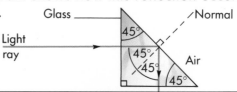

Back   20 cm   Front   Pinhole
10 cm
10 cm
Greaseproof paper   Black paper
Box made of black card

(a) How big is the image of the lamp?

(b) How big would the image be if the lamp was 100 cm from the camera?

(c) How would the picture differ if the pin-hole was enlarged?

PRISMS

Prisms are triangular blocks of glass or plastic. A prism can be used to reflect light rays like a plane mirror. The diagram shows how this reflection occurs. It is called **total internal reflection**.

Glass   Normal
Light ray   45°
45°   Air
45°
45°

ACTION!

In camera

Look at the diagram of a prismatic camera.

Explain using the diagrams how this type of camera operates.

Can you suggest any advantage of this type of camera.

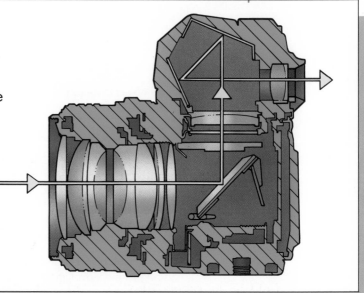

146

A triangular prism can also be used to split white light up into its constituent colours. The band of colours formed on the screen is called the **spectrum** of white light. When a ray of whitelight enters a prism, the light is split up or **dispersed** because the different colours are refracted by different amounts.

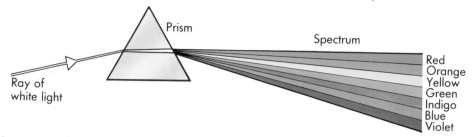

The same dispersion of light occurs in a rainbow. In this case rain droplets disperse the light.

If the separate colours are mixed, white light is produced. In fact only three colours have to be mixed to produce white light. These are red, green and blue. They are called **primary light colours**. If a pair of primary light colours are mixed, a **secondary light colour** is formed. For example: red and green produce yellow; and blue and red produce magenta/cyan.

Look at the shape below which summarizes the colour of light produced when different colours are mixed. White light is produced in the middle where red, blue and green are mixed.

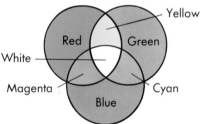

If white light is passed through a perfect red filter, only red light passes through the filter. If white light is passed through a red filter and then a blue filter, no light will pass through. Why should this be so?

THE EYE

An eye is a very complicated optical instrument. It is able to detect shapes and colour. Look at the diagram of an eye on page 148.

An eye contains a convex lens which focuses the light rays on to the retina. Cells in the retina are stimulated by light. Messages are sent from them to the brain via the optic nerve. The ciliary muscles of the eye can alter the shape of the

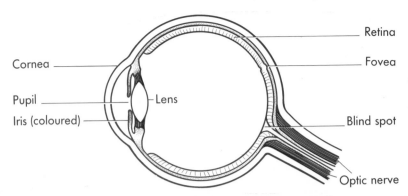

Cornea
Pupil
Iris (coloured)
Lens
Retina
Fovea
Blind spot
Optic nerve

lens, making it thicker or thinner. This enables the eye to focus clearly on objects at different distances. This is called **accommodation**. When the ciliary muscles contract, they squash the lens decreasing the focal length. This makes close objects come into focus. An average young person can focus on near objects that are about 25 cm away, without straining. This distance is called the **near point**. If the muscles relax, the lens becomes thinner. The focal length increases. The eye can now focus on distant objects. A normal eye will have a **far point** at the horizon (i.e. infinity).

The image produced on the retina is upside down and smaller than the object.

The iris adjusts the 'hole' in the eye, called the pupil, allowing different amounts of light into the eye.

There are no nerve endings where the optic nerve leaves the eye. This produces a **blind spot** where an image cannot be formed.

Various eye defects can occur when different parts of the eye do not function properly. Usually these can be corrected by additional lenses such as those in spectacles or contact lenses.

Short sight

A short-sighted person can focus the light from near objects and form a sharp image on the retina. However, with distant objects the lens is too strong. It converges the light rays to form an image in front of the retina.

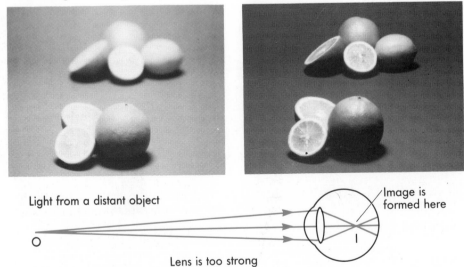

Light from a distant object

O

Image is formed here

Lens is too strong

Short sight can be corrected by using a concave lens in front of the eye.

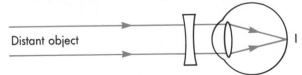

Distant object

I

Long sight

If the ciliary muscles are too weak, they will not be able to squeeze the lens sufficiently to form a sharp image. This is something which happens as people get older. It is then difficult to focus on nearby objects although far away objects will appear sharp. This is called **long sight**.

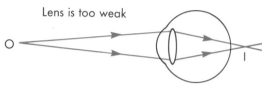

It can be corrected using a convex lens.

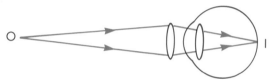

Find out about astigmatism.

ACTION!

The eyes have it

Some animals have two eyes on the side of their head rather than at the front of the head as we do.

What are the advantages of having eyes arranged:
(a) on the side of the head; (b) at the front of the head?

OPTICAL INSTRUMENTS

When looking at a map, you can use a lens as a magnifying glass.

Your eye should be as close to the lens as possible.

The magnifying glass produces a magnified virtual image.

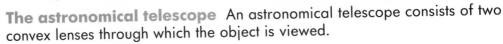

Other optical instruments are more complicated:

The astronomical telescope An astronomical telescope consists of two convex lenses through which the object is viewed.

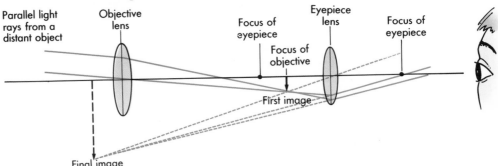

Parallel light rays from a distant object · Objective lens · Focus of eyepiece · Focus of objective · Eyepiece lens · Focus of eyepiece · First image · Final image

The lens closest to the object is called the **objective** lens. This is a weak converging lens. The other lens, closer to the eye, is called the **eyepiece** lens. This is a stronger converging lens. In a powerful telescope, the weak objective lens will make a real image a long way from the lens. The telescope must be a long instrument.

The microscope This also consists of two convex lenses. However, both convex lenses are strong in this case.

The camera The lens in a camera is a convex lens. When the shutter is opened, light rays from the object are focused to form an image on the film. The image produced is real, upside down and smaller than the object.

In a complicated camera, the lens can be adjusted so that the image formed on the film is sharp and in focus.

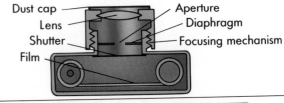

Dust cap · Lens · Shutter · Film · Aperture · Diaphragm · Focusing mechanism

ACTION!

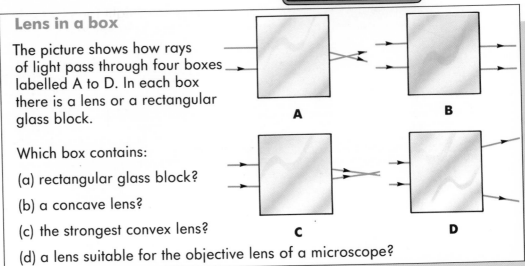

Lens in a box

The picture shows how rays of light pass through four boxes labelled A to D. In each box there is a lens or a rectangular glass block.

Which box contains:

(a) rectangular glass block?

(b) a concave lens?

(c) the strongest convex lens?

(d) a lens suitable for the objective lens of a microscope?

Fibre optics

A beam of light can pass through a fine thread of glass. This acts as a guide and the light can only come out at the other end of the thread.

The fibre is very thin; thin enough to pass through the eye of a needle. The fibre can be curved to go around corners.

Light can be made to change direction but the diagram shows that the light still travels in straight lines.

Fibre optics is expanding rapidly. Light guides can be made small enough to be used in the body to give pictures of vital organs working (*see* photograph).

Modern telephone systems use fibre optics in place of copper wire. The messages are flashes of light rather than electrical systems. The advantages of this system are:

1 More conversations can be carried along a light guide than a copper wire;
2 The light guide is thinner than a copper wire;
3 There is less interference of one message with another;
4 Fewer parts to go wrong;
5 Less electrical energy is required.

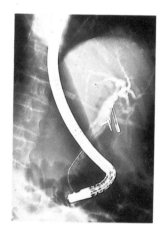

OTHER FORMS OF ELECTROMAGNETIC RADIATION

Light is only one form of **electromagnetic radiation**. All electromagnetic radiation consist of waves such as the simple transverse wave below. In the diagram the wave moves up and down but the energy moves across from right to left.

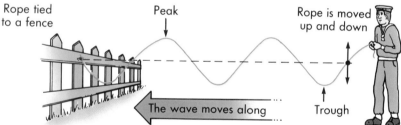

Look at the electromagnetic spectrum of which light forms a part. The diagram also shows some uses of these different forms of radiation.

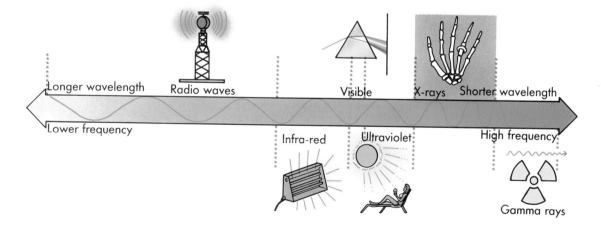

UNIT 16

THE EARTH IN SPACE

You should already know that you should never look directly at the Sun as this could damage your eyes. You should also realize that the Sun moves across the sky during the day: rising in the east and setting in the west. The Sun is higher in the sky in the Summer than in the Winter.

You should be aware that the Earth is a large ball which is constantly turning. Night occurs when our part of the Earth is turned away from the Sun. The length of daylight changes throughout the year.

THE SUN

The Sun is the centre of our **solar system**. The Sun is only one of billions of stars in the Milky Way.

The Sun is extremely hot and is radiating energy. This energy enables the Earth to function. The energy of the Sun is produced by atomic **fusion**. Fusion involves joining small atoms together to form larger atoms. There is a large amount of energy given out in this process. The Sun is a mixture of hydrogen and helium gases. The fusion of hydrogen atoms to form helium atoms produces energy. The temperature of the Sun at its centre has been estimated at 20 000 000°C and at its surface about 6000°C.

The Sun is about half-way through its life of about 9600 million years. Look at the life cycle of a star such as the Sun.

Giant cloud of dust and gas — Compressed by its own gravity — Reactions start which release energy — The star expands — The star collapses to form a white dwarf star — All reactions have stopped: no energy is en[...] — A black hole is forme[...]

Planet	Diameter Earth =1	Mass Earth =1	Surface gravity Earth =1	Density (kg/m³)	Average distance from sun Sun–Earth =1	Period of orbit (years)	Number of moons
Earth	1.00	1.00	1.00	5500	1.0	1.0	1
Jupiter	11.18	317.00	2.60	1300	5.2	11.9	16
Mars	0.53	0.10	0.40	4000	1.5	1.9	2
Mercury	0.40	0.06	0.40	5400	0.4	0.2	0
Neptune	3.93	17.20	1.20	2300	30.1	164.8	2
Pluto	0.31	0.0025	0.20	400	39.4	247.7	1
Saturn	9.42	95.00	1.10	700	9.5	29.5	15
Uranus	3.84	14.50	0.90	1600	19.2	84.0	5
Venus	0.95	0.80	0.90	5200	0.7	0.6	0

Moving around the Sun are the **planets**. Each planet moves along a path called an **orbit**. The orbits are not circular, but oval-shaped. Can you suggest a reason for this?

The table on page 152 gives some information about the planets. The drawing below shows the relative distance between them.

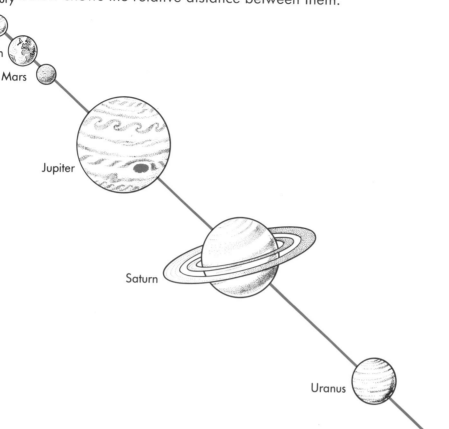

Sun

Mercury

Venus

Earth

Mars

Jupiter

Saturn

Uranus

Neptune

Pluto

ACTION!

Ordering orbits

Use the table on page 152 to answer the following questions:

(a) Which of the planets:
   (i) takes the shortest time to orbit the Sun? (ii) have no moons?
   (iii) has the greatest density? (iv) has the largest diameter?

(b) Arrange the planets in order of distance from the Sun starting with the one which is closest to the Sun.

(c) How do the diameters and densities of planets close to the Sun differ from those of planets furthest from the Sun?

Planet X

Using the table on page 152 identify the planets described on the following computer screens:

PLANET A
Length of day: 10 hours
Length of year: 12 years
Much larger than Earth
Number of satellites: 16
Temperature of planet: −120°C

PLANET B
Length of day: 4 months
Length of year: 7 months
Smaller than Earth
Number of satellites: 0
Temperature of planet: 480°C

PLANET C
Length of year: 12 months
Type of surface: water and rocks
Number of satellites: 1
Temperature of planet: 20°C

Plutonic poser

Scientists studying the orbit of the planet Pluto have noticed that its orbit alters from time to time and they do not know why. Can you suggest any explanation for this?

It is difficult to appreciate the size of these distances. The following example may help. Imagine the Earth is a small marble and the Sun is a beach-ball about 1 m in diameter. In this model of the Earth and the Sun, the marble and the beachball would be 100 m apart. The Moon, which orbits the Earth, is represented by a ball-bearing (diameter 2.5 mm) about 25 cm from the marble.

Did you know?

The Earth as the centre of the Universe

Read the first chapter of the book of Genesis in the Bible. This contains a story of how the universe was created. Notice that the writer did not understand the way the planets orbited the Sun. The Earth is described as the centre of the Universe. According to the theory developed by Ptolemy, the Earth was stationary. He thought that the Moon, planets and stars moved round the Earth in circular orbits.

This theory was accepted by the Church because it left room outside the fixed stars for heaven and hell.

This theory was accepted until 1543 when Nicholas Copernicus dared to suggest that the planets, including the Earth, orbited the Sun.

Nicholas Copernicus was a Polish man who was asked by the Catholic Church to look at ways in which the calendar could be improved. After looking at the paths of the planets in the sky he concluded that the Earth, and the other planets, were moving round the Sun. Copernicus realized that his conclusions went against the teaching of the Church. So, he did not publish his conclusions until the last year of his life.

Johannes Kepler (1571-1630) worked out the orbits of the planets in more detail. Galileo Galilei (1564-1642) built the most powerful telescope at the time. This could magnify up to thirty times. He was able to see the planet Jupiter and the four moons which move round it. The fact that the moons moved around Jupiter showed that everything could not rotate around the Earth. Galileo was threatened with torture by the Church. He was forced to sign a document saying that he did not believe that the Sun was the centre of the Universe.

Shortly after Galileo's time Isaac Newton worked out the orbits of the planets using the laws of gravity. This finally convinced people that the Sun was the centre of the solar system (*see* photograph opposite).

(a) Why did the Church accept the Ptolemy's theory and object to the theory that the Earth was not the centre of the Universe?

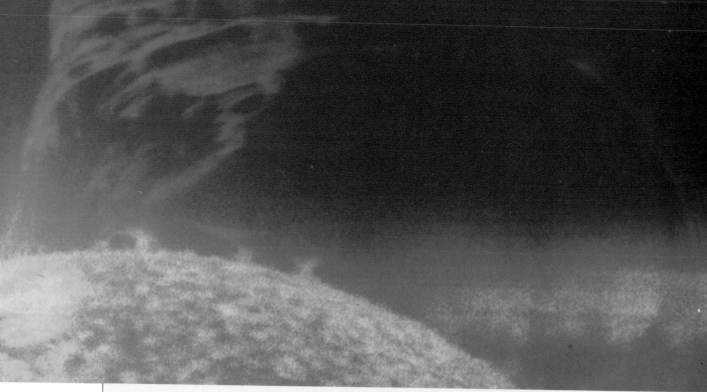

(b) Newton's law of gravity explains many of the movements of objects in space. It states that all objects are attracted to one another by a force of gravity. The size of the force depends upon how close the objects are and on how large they are. The force is greatest when large masses are near to one another.

Use this information to explain why:

 (i) the path of a spacecraft changes as it comes close to the Moon but the path of the Moon does not change;

(ii) the paths of planets can be affected by the movement of other planets.

Did you know?

The planet Saturn

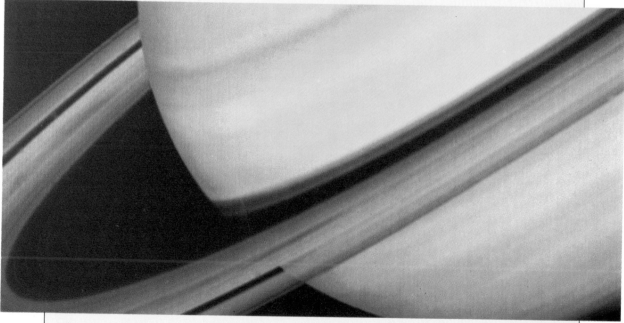

The planet Saturn was first seen by Galileo through his telescope in July 1610. The illustration shows what he saw.

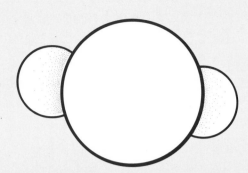

From this observation, Galileo concluded that Saturn was a 'triple planet'.

He described this as a large planet with two moon-like bodies very close to it.

When he looked again in 1612, he could see only a single planet. The two moons had disappeared. In 1613 the 'triple planet' was visible again.

In 1656, the Dutch astronomer Christiaan Huygens was the first to see that Saturn has rings around it. It was not clear until 1875 that these rings are made up of millions of tiny particles. They are probably lumps of frozen gas.

(a) Suggest a reason why Galileo was unable to distinguish rings around Saturn whereas Huygens could.

(b) Can you explain why Galileo's view of Saturn was different in 1612.

Night and day

As the Earth travels around the Sun it also spins. Imagine a long rod pushed through the North Pole and coming out through the South Pole. This rod is called the Earth's **axis**. The Earth makes one complete spin on its axis each day (24 hours). As the Earth spins, one half faces the Sun, (day time). The other half faces away from the Sun, (night time).

Looking down on the North Pole, the Earth spins anticlockwise

As the Earth spins in this direction, the Sun appears to move across the sky from east to west.

In fact, the Sun remains still and the Earth is moving.

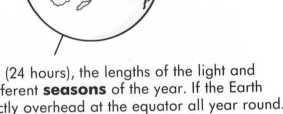

Although every day is the same length (24 hours), the lengths of the light and dark periods change. This leads to different **seasons** of the year. If the Earth were not tilted, the Sun would be directly overhead at the equator all year round. In this situation, there would be no changes in weather patterns during the year.

The seasons are formed because there is a 23° tilt of the Earth's axis. As the Earth travels around the Sun, different areas of the Earth's surface are directed towards the Sun at different times of the year.

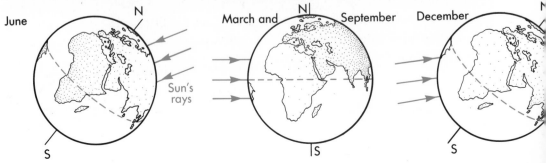

Two imaginary lines, parallel to the equator (called the Tropic of Cancer and the Tropic of Capricorn) are the limits at which the Sun appears vertically overhead at some time in the year.

When it is summer in the Northern Hemisphere, it is winter in the Southern Hemisphere and vice versa. The Sun is directly above the equator in spring and autumn.

Months and years

Each planet is a different distance from the Sun. The time taken for each planet to go round the Sun is different in each case. The Earth takes 365 days and 6 hours to orbit the Sun. This is called one year. Years are taken as 365 days with a **leap year** every four years. This is a year of 366 days to catch up on the spare 6 hours each normal year. In a leap year there is an extra day in February.

The moon is a satellite of the Earth.

It takes 28 days to orbit the Earth.

This is called a **lunar month**.

Our view of the Moon varies at different times of the month depending upon the illumination from the Sun.

The diagram shows the different **phases** of the Moon.

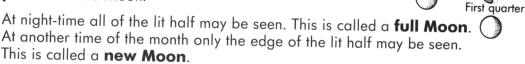

At night-time all of the lit half may be seen. This is called a **full Moon**. At another time of the month only the edge of the lit half may be seen. This is called a **new Moon**.

The new Moon looks dark because the lit side is facing away from the Earth, so that only the side that is in darkness is seen. It is faintly visible because the Earth reflects sunlight onto the Moon.

Did you know?

Does the same side of the Moon always face the Earth?

Imagine that your friend is in a revolving chair and that you are walking round the chair in a large circle. As you move round, your friend turns the chair so that he or she is always facing you. When you have walked round the circle once, will you ever have seen your friend's back. The answer is no!

A similar situation applies to the Moon. We only ever see one side of the Moon. This was confirmed in 1959, when the Russian spacecraft LUNA 3 photographed the back of the Moon.

Lunar and solar eclipses

When the Earth and the Moon are in certain positions, they block out the light from the Sun. We call this an **eclipse**. There are two types of eclipse: lunar (moon) eclipse; and solar eclipses.

Lunar eclipses are more common than solar eclipses. Look at the diagram of a lunar eclipse. The Earth stops light from the Sun from illuminating the Moon. The Moon is in the Earth's shadow and it cannot be seen from the Earth.

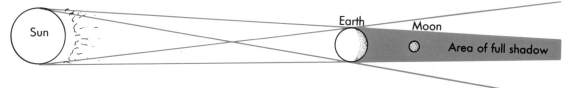

The diagram below shows an eclipse of the Sun (solar eclipse). The Moon passes between the Sun and the Earth and so produces a shadow on the Earth. When this happens, the Sun cannot be seen from those parts of the Earth in full shadow from the Moon.

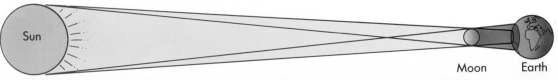

Sun   Moon   Earth

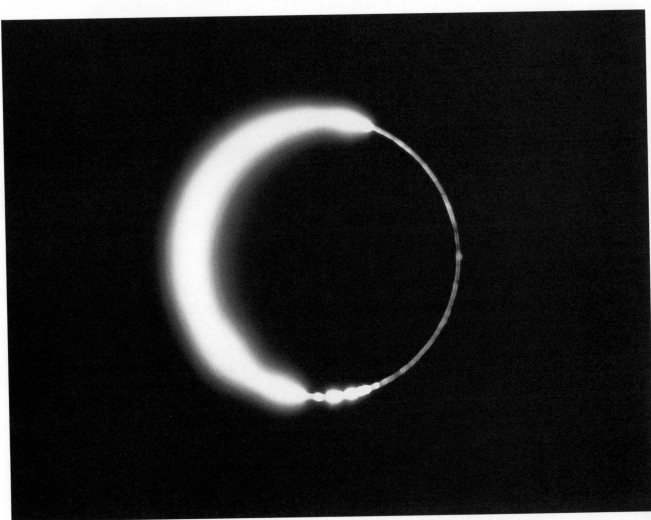

THE UNIVERSE

Much is known about the Sun and the planets in our solar system. However, you only have to look at the sky on a starry night to appreciate that there are a vast number of stars in the sky. Each star seems no bigger than a twinkling speck. Yet, all of these stars are much bigger than the Earth and some are millions of times bigger.

Distances in space are so vast that it is very difficult for them to be appreciated. Light waves and radio waves travel at 298 000 km per second. So, a radio signal from the Earth would take: 1.25 seconds to reach the moon; 8 minutes to reach the sun; and 4 years to reach Proxima Centauri, our nearest star.

Stars exist in **galaxies** and a number of galaxies together is called a **group**. The Local Group in which the Earth exists, contains 27 galaxies. A distant galaxy in our group is called the Andromeda Spiral. It is just visible from the Earth with the naked eye. A radio wave signal from the Earth would take 2.2 million years to reach Andromeda Spiral (*see* opposite).

Did you know?

The Earth is a sphere; not a disc

In 340 BC, the Greek scientist Aristotle published a book called *On the Heavens* In this book he argued that the Earth was a sphere and not a disc. He used two observations:

1 When a ship is sailing towards you from some distance away, you see the mast before you see the hull. He was appreciating the curvature of the Earth.

2 During an eclipse of the Moon, a shadow is formed on the Moon. The shadow caused by the Earth is always round.

What shape would the shadow have on the Moon if the Earth was a disc?

ACTION!

The sky's the limit?

Why can scientists not be certain when they observe the sky that the Andromeda Spiral exists today?

These distances are so immense that they are difficult to comprehend. If the Sun was reduced to the size of a hydrogen atom, (the smallest atom known), the galaxy containing the solar system would have a diameter of 69 m on this scale. Andromeda Spiral would be about 1.6 km away and the edge of the Universe about 3200 km away.

Did you know?

History of the Universe

In 1929, Edwin Hubble observed that distant galaxies were moving away from one another. In other words, the Universe is expanding.

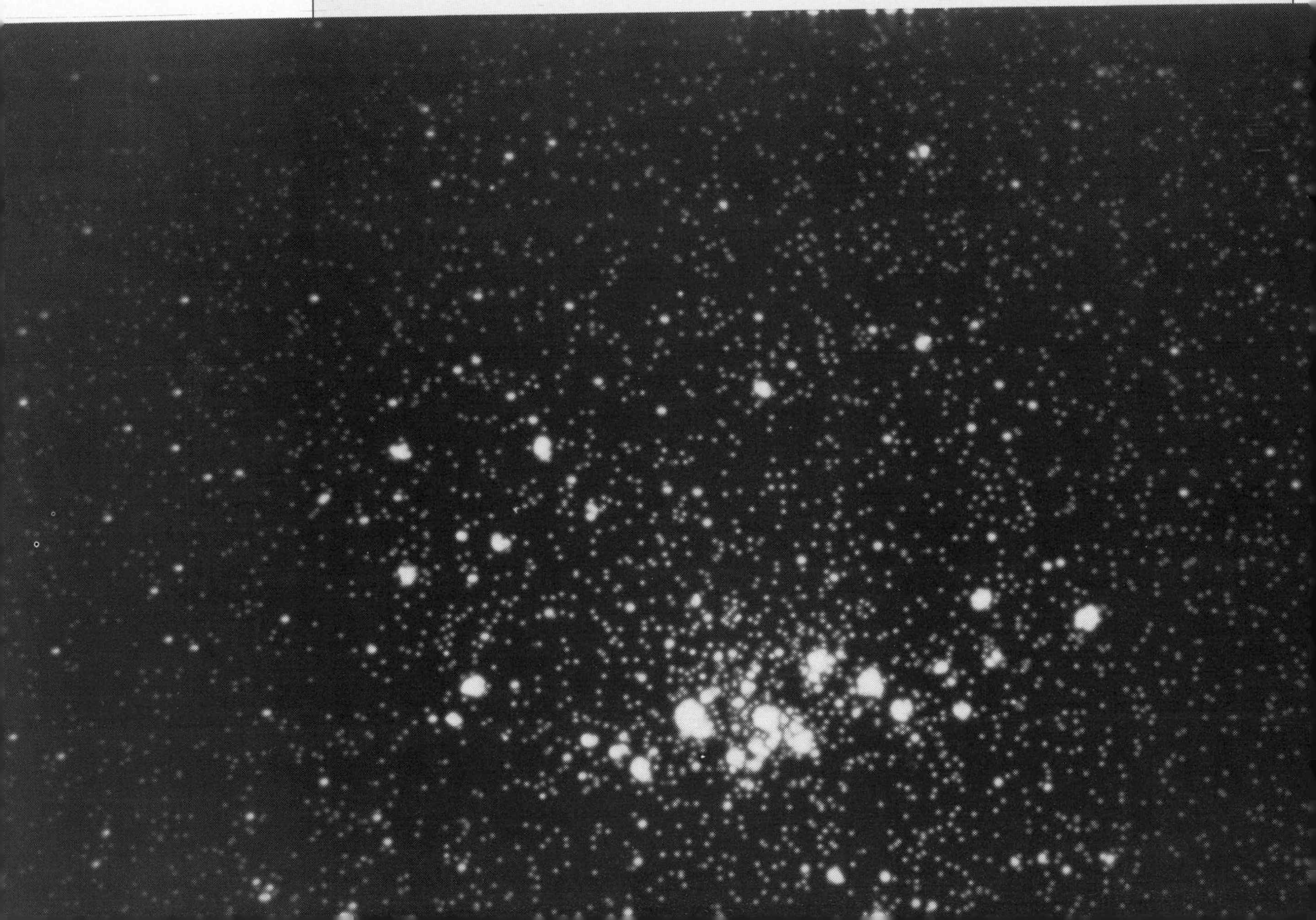

He worked backwards and showed that between 10 000 and 20 000 million years all of the galaxies would have been at the same point. Then the density and temperature of the mass would both have been extremely high. He suggested that the Universe had been created at the time by a 'big bang'. This theory was accepted by the Catholic Church in 1951.

(a) What venture in 1990 was launched and named after Hubble?

(b) Suggest reasons why the Catholic Church accepted the 'big bang' theory quite readily.

PROJECT

Staring into space You can learn a great deal about stars, planets and other heavenly bodies by looking carefully at the sky. You do not need a telescope. A pair of binoculars give you a good starting point. With them you can see the craters on the Moon, the phases of Venus, the moons of Jupiter and a multitude of stars.

Never look at the Sun directly especially through binoculars or a telescope.

You will also need a compass so you can find north, south, east and west.

Every month in many newspapers, star maps are published which tell you where to look in the sky for particular features. You can find out more by contacting the Junior Astronomical Society (the address is on page 176).

If you watch the sky for some time at night, you will notice that the pattern of stars does not change but revolves around the Pole Star once every 24 hours. The stars move from east to west.

Find out the shape of the following groups of stars: the Plough; Cassiopeia; Cygnus (the Swan).

Look to the north and find the bright star we call the Pole Star. This does not move. Try to identify the groups of stars above you. Draw a star chart to show how these groups fit around the Pole Star. Why can this activity only be done in the Northern Hemisphere.

The Moon travels across the star pattern. Sometimes it is among the stars and sometimes it is not. Over a number of hours, observe the position of the Moon among the stars. In the course of a night, the Moon also sweeps across the sky from east to west but not quite so fast. If you watch carefully, you will see that the Moon lags further behind the stars as the month goes on.

You will be able to see some of the planets in the sky. You might want to visit a planetarium or museum to see displays of the stars and planets.

GLOSSARY OF TERMS

acceleration Increase in velocity (or speed) in a given time.

accommodation Ability of the eye to focus automatically.

acid Substance that dissolves in water to form a solution with a pH of less than 7. An acid usually contains hydrogen which can be replaced by a metal to form a salt.

algae Simple plants which may be single-celled or much larger, (e.g. seaweeds).

alkali Base that dissolves in water to form a solution with a pH above 7. Alkalis are neutralized by acids to form salts.

alloy Metal made by mixing two or more metals together, (e.g. brass is an alloy of copper and zinc).

alternating current (a.c.) Current which keeps changing direction around an electrical circuit.

ammeter Instrument for measuring current flow.

amoeba Single-celled animal found in the mud at the bottom of ponds.

amphibians Animals with backbones which have moist skins without scales. They live in and out of water, (e.g. frogs).

anhydrous Substance without water. Often used to describe substances which have lost water of crystallization.

anion Negatively charged ion which moves towards the anode during electrolysis, (e.g. chloride ion, Cl^-).

anode Positively charged electrode in electrolysis.

anther Male part of a flower which produces pollen.

antibodies Chemicals made by the body to kill germs.

aqueous solution Solution made by dissolving a substance in water.

arachnids Animals with four pairs of jointed legs, (e.g. spiders).

arteries Blood vessels which carry blood away from the heart.

arthropods Animals with no backbone, but with jointed legs.

asexual reproduction Reproduction that requires only one parent, (e.g. taking a geranium cutting).

atom Smallest part of an element that can exist that has all the properties of that element.

atria Upper two chambers of the heart.

bacteria Microscopic organisms which may cause decay.

barometer Instrument for measuring atmospheric pressure.

base Substance which reacts with an acid to form a salt and water only, (e.g. metal oxides).

boiling Liquid turns rapidly to a vapour at a fixed temperature called the boiling point.

bones Hard substance which makes up the skeleton. Bones are made up of protein fibres and mineral salts.

capillary Smallest blood vessels in the body which link the arteries and the veins.

carbohydrates Compounds of carbon, hydrogen and oxygen which contain twice as much hydrogen as oxygen, (e.g. glucose $C_6H_{12}O_6$).

carnivore Animal which eats flesh, (e.g. tiger).

carpels Female reproductive organs in a flower.

catalyst Substance which alters the rate of a chemical reaction but is not itself used up in the reaction.

cathode Negatively charged electrode in electrolysis.

cation Positively charged ion which moves towards the cathode in electrolysis, (e.g. hydrogen ion, H^+).

cell membrane Thin skin which surrounds an animal cell. Plant cells have a cellulose cell wall that surrounds the cell membrane.

cells Tiny units which make up living matter.

charges Either positive or negative, they exert forces on each other: like charges repel; unlike charges attract.

chemical change Change which results in the formation of new substances.

chlorophyll Green pigment in plants which absorbs light energy to start the photosynthesis reaction.

cholesterol Fatty material found in some foods which can be deposited in the arteries.

chromatography Way of separating mixtures, especially of coloured substances, by letting them spread across a filter paper or through an absorbant powder.

ciliary muscles Muscles which control the shape of the eye lens and keep it in focus.

classification Sorting out of information into groups; usually referring to living things.

combination Atoms of different elements join together to form a compound.

combustion Burning of a substance in oxygen.

community Group of organisms which live together in a habitat.

components Parts that go up to make an electrical circuit.

compound Substance formed by joining atoms of different elements together. The properties of a compound are different from the elements that make it up. The proportions of the different elements in a particular compound are fixed.

concave Lens or mirror which curves inwards (i.e. cave-like).

condensation Occurs when a vapour turns to a liquid on cooling. Heat is given out during the change. Condensation is the opposite of evaporation.

conductor Allows electricity to pass through it (electrical conductor) or heat to pass through it (heat conductor). Metals are good conductors of heat and electricity. Carbon (graphite) is a good electrical conductor but a poor heat conductor.

convex Lens or mirror which curves outwards.

cornea Tough, outer coating on the front of the eye.

corrosion Wearing away of the surface of a metal by chemical attack, (e.g. rusting of iron and steel).

crest Top of a wave. The lowest point on a wave is called the **trough**.

critical angle Angle at which the light is all internally reflected and not refracted.

crustaceans Arthropods which have two pairs of antennae, from nine to fifteen pairs of legs and often a hard, chalky shell.

crystal A piece of a substance that has a definite regular shape. Crystals of the same substance have the same shape. Slow crystallization will produce larger crystals.

current Flow of electricity.

cytoplasm Jelly-like material which makes up the contents of a cell.

deceleration Decrease in velocity (or speed) over a given time, sometimes called negative acceleration.

decomposers Bacteria and fungi which break down the bodies of dead plants and animals and the waste products of living organisms.

decomposition Chemical reaction that results in the breaking down of substances into simpler ones.

density Mass of a particular volume of a substance. It is expressed as kg/m^3 or g/cm^3.

detergents Cleaning agents of two main types: soaps and soapless detergents.

diffuse reflection Light is reflected in all directions from a surface.

diffusion Movement of particles from where they are highly concentrated to where they are less concentrated.

digestion Breaking down of large, insoluble food molecules into smaller, soluble molecules.

direct current (d.c.) Current that travels in one direction around an electrical circuit.

dispersion Splitting of white light into the different colours of the spectrum.

dissolving Occurs when a substance is added to water and disappears from view when stirred. The substance is still there and can be recovered by evaporation.

distillation Way of purifying a liquid or obtaining the solvent from a solution. The liquid is vaporized and the vapour condensed to reform the liquid. The condensed liquid is called the distillate.

ductile *See* metal.

ecology Scientific study of the way living things relate to one another and their environment.

efficiency Measure of the amount of energy wasted by a machine or process.

electrode Conducting rod, or plate, which carries electricity in, or out, of an electrolyte during electrolysis.

electrolysis Decomposition of an electrolyte, either molten or in aqueous solution, using electricity. The **electrolyte** is usually an acid, base, alkali or salt.

electrostatics Study of electric charges.

element Single pure substance that cannot be split up into any simpler chemical.

endothermic reaction Reaction that takes in heat.

energy Quantity that enables objects to do something or make something happen. Energy can be kinetic energy (due to movement) or potential energy (due to position).

environment Surroundings in which animals and plants live.

enzyme Protein which acts as a biological catalyst.

equilibrium Position when a set of forces is completely balanced.

erosion Wearing away of materials, (e.g. rocks).

evaporation Process by which a liquid changes to its vapour. This happens at a temperature below the liquids boiling point but is fastest when it is boiling.

exothermic reaction Reaction that gives out heat, (e.g. the burning of coal).

eyepiece lens Lens in an optical instrument, such as a microscope, closest to the eye of the operator. The lens closest to the object is called the **objective lens**.

far point Furthest point which can be seen clearly.

fermentation Enzymes in yeast convert glucose into ethanol an carbon dioxide.

fertilization Joining together of a male sex cell with a female sex cell.

filtrate Liquid that comes through the filter paper during filtration.

filtration (or filtering) Method of separating a solid from a liquid. The solid collects on the filter paper and the liquid runs through.

flammable Substance which catches fire easily (e.g. petrol).

food chain Series of organisms (producers and consumers), starting with a green plant, which eat and are eaten by each other.

food web Series of interconnected food chains.

force Pushes, pulls, or turns which are measured in newtons (N).

fossil Preserved remains of organisms which lived millions of years ago.

fractional distillation Method of separating a mixture of different liquids using differences in their boiling points.

freezing A liquid changes to a solid at the freezing point. A pure substance will have a definite freezing point.

frequency Number of events in one second.

fuel Substance that burns easily to produce heat and light. A **fossil fuel** is present in the Earth in limited amounts only, and cannot be readily replaced, (e.g. coal, petroleum).

genus A group of similar species.

germination Beginning of growth of a seed.

germs Microbes which cause disease.

gravity Attractive force of a body for a nearby object. The larger the body the stronger will be the **gravitational force**.

habitat Part of an environment where a community of organisms live, (e.g. a pond).

herbivore Animals which only eats plants, (e.g. cow).

hermaphrodite Living organisms which develops both male and female reproductive organs, (e.g. earthworm).

hormone Chemical which regulates certain processes in the body.

humus Broken down remains of plants and animals found in the soil.

hydrolysis Splitting up of a compound using water.

igneous Rocks that have cooled and solidified from molten rock, (e.g. granite).

immiscible Two liquids that do not mix, (e.g. oil and water).

immunity Ability of the body to resist infection by disease-causing organisms.

incident ray Light ray which travels *into* an optical instrument.

indicator Chemical that can distinguish between an alkali and an acid by changing colour, (e.g. litmus is red in acids and blue in alkalis).

inertia Reluctance of a mass to move, or have its movement changed.

insecticides Poisonous chemicals which can kill insects. They can often affect the environment.

insoluble Describes a substance that will not dissolve in a particular solvent.

insulator Substance which does not conduct electricity, (e.g. rubber or plastic).

internal reflection Reflection that takes place on the inside surface of the material. The incident ray strikes the surface with an angle greater than the critical angle.

ion Positively, or negatively, charged particle formed when an atom, or group of atoms, lose or gain electrons.

iris Coloured, circular portion at the front of the eye. It adjusts to alter the amount of light entering the eye.

key Series of written instructions used to identify living organisms.

kinetic energy *See* energy.

laterally inverted The image is turned the wrong way round.

lens An optical device made of glass, or plastic, which bends light.

long-sighted Person who is not able to see close objects clearly without correction. It is corrected with a concave lens.

malleable *See* metal.

mass The amount of matter in a material. It is important to distinguish mass from weight.

melt A solid changes to a liquid at the melting point.

metal An element that is: shiny; conducts heat and electricity; can be beaten into thin sheets (**malleable**); or drawn in wires (**ductile**), is probably a metal.

metamorphic Igneous or sedimentary rocks were thoroughly altered by heat and/or pressure within the crust of the Earth without melting to form these rocks, (e.g. marble).

microbes Microscopic plants and animals including bacteria and viruses. Bacteria can be killed by antibiotics, such as penicillin, but viruses cannot.

mineral Naturally occurring substance from which rocks are made.

mixture Substance made by just mixing other substances together. This is not a chemical reaction.

molecule Smallest part of a diatomic element (e.g. H_2) or compound that can exist on its own.

near point Closest point which can be seen clearly.

nectar Sugary liquid produced by plants to attract insects for pollination.

neutralization A reaction where an acid is cancelled out by a base or alkali.

omnivores Animals which eat plants and animals, (e.g. humans).

organ Collection of tissues grouped together to make a structure with a specific job, (e.g. the heart).

osmosis Movement of water from a dilute solution to a more concentrated solution through a semi-permeable membrane, (i.e. movement from a high concentration of water to a low concentration of water).

ovaries Female sex organs which produce ova.

ovum Female sex cell, sometimes called an egg.

oxidation Reaction where a substance gains oxygen or loses hydrogen.

pH Measure of the acidity, or alkalinity, of a substance. A substance with a pH of 7 is neutral.

photosynthesis Process by which green plants use the energy from the sun to fuel the reaction that builds up their food.

plasma Liquid part of the blood.

pollution Substances in the environment which are harmful to living things.

potential energy *See* energy.

power Rate of using energy.

precipitate Insoluble substance formed in a chemical reaction which causes the solution to become cloudy.

predator Animal which catches and eats other animals, (prey).

pressure Force spread over an area.

properties Description of a substance and how it behaves. Physical properties include density and melting point. Chemical properties describe chemical changes.

proteins Important body-building foods including meat, fish and eggs.

pupil Hole through which light passes in the center of the iris of an eye.

random motion Uncontrolled motion which cannot be predicted.

reactant Chemical substance which takes part in a chemical reaction.

real image Image which can be projected onto a screen.

red blood cells Contain haemoglobin and carry oxygen around the body via the blood system.

redox reaction Reaction where both *red*uction and *ox*idation occur.

reduction Opposite of oxidation: a reaction where oxygen is lost and hydrogen is gained.

refraction Bending of light when it passes from one material to another.

reproduction Process by which living organisms produce more of their own species.

residue Insoluble substance left on the filter paper during filtration.

resistance Slowing down or stopping of the flow of electricity in a circuit by a material, measured in ohms (Ω).

resonance Object vibration in sympathy with something else.

respiration Chemical process in living things which releases energy from food.

retina Light-sensitive coating on the back of the inside of the eye.

salt Substance which is formed as a product of a neutralization. A salt is the product when hydrogen in an acid is replaced by a metal.

scalar Measurement where size, not direction, is important, (e.g. speed).

sedimentary Rocks composed of compacted fragments of older rocks, and other minerals, which have accumulated on the floor of a sea or lake, (e.g. sandstone).

sepals Leaf-like structures which enclose and surround an unopened flower bud.

sex cells Cells which join together during fertilization.

sex organs Parts of the body which produce sex cells.

short-sight Person unable to see faraway objects clearly without correction. It is corrected using a convex lens.

soil Complex mixture of rock fragments, mineral salts, humus and living organisms in which most of the world's plants grow.

solubility Grams of a solute that will dissolve in 100 g of solvent at a particular temperature.

solute Substance that dissolves in a solvent to form a solution.

solvent Liquid in which the solute dissolves.

species Group of organisms which can breed with each other to produce fertile offspring.

stamens Male sex organs in flowering plants.

stigma Part of the carpel to which the pollen grains stick during pollination.

synthesis Formation of a compound from the elements that make it up.

tissues Groups of similar cells which perform a similar task.

toxin Poisonous substance, often produced by bacteria.

umbra Area of complete shadow. Partial shadow is called **penumbra**.

vacuole Fluid-filled space inside a cell.

vapour Gas that will condense to a liquid on cooling to room temperature.

vector Measurement where size and direction are both important, (e.g. velocity).

veins Vessels in animals which return blood to the heart.

virtual image Image which cannot be projected on to a screen.

volatile Describes a liquid which turns easily to a vapour, (e.g. petrol).

voltage Indicates the energy value of an electric current. It is measured in volts (V).

water of crystallization Definite amount of water bound up in a crystal, (e.g. hydrous copper II sulphate, $CuSO_4.5H_2O$).

wavelength Distance between two identical points on a wave.

white blood cells Blood cells which help the body fight disease. They do not contain haemoglobin.

work Work is done when a force moves.

yeast Single-celled fungus which releases carbon dioxide when it respires.

ANSWERS

Unit 2

Puzzle: Cartoon differences
(a) Changed head
(b) Straw
(c) Beetle
(d) Patch on sleeve
(e) Mouse
(f) Knot in
(g) Label on bottle: top shelf
(h) Lid on jar: second shelf
(i) Missing bubble: extreme left
(j) Extra spider's leg

Nymph spotting: Action
The similarities include:
one pair of antennae;
six legs;
separate head, thorax and abdomen, etc.

Spot the difference: Action

1 (a)

A	B
6 legs	10 legs
wings	no wings
no claws	claws
separate parts	no separate parts

(b)

C	D
no legs	legs
no antennae	antennae
clitellum/saddle	no clitellum/saddle

Spine time: Action
Vertebrates: fish; frog; monkey; pigeon

Invertebrates: earthworm; jellyfish

In hot and cold blood: Action
Warm-blooded: Penguin; mouse
Cold-blooded: turtle; lizard; snake; fish

Class creatures: Action
Amphibian, fish, reptile, bird and mammal

Marianne North: Did you know?

Photography; new printing methods; methods of communication (telephone, television, fax etc); more freedom of movement, especially for women.

The key to leaves: Action

A Oak; **B** Lilac; **C** Fraxinus; **D** Platanus; **E** Alnus; **F** Aesculus

Name that fish: Action

There are many possibilities. This is one.

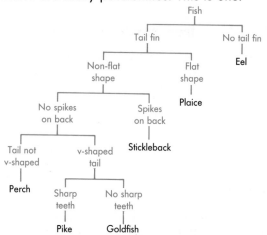

Identify the creatures: Action

A Insect; **B** Myriapod; **C** Crustacean; **D** Arachnid;

Finding a niche: Action
They have flattened bodies; claws to hold onto stones; and gills to enable them to get oxygen from the water.

Surveying the sites: Action

(a) Aphids
(b)

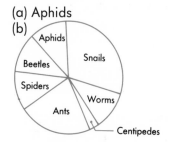

(c) Centipedes occurred in fewer numbers: therefore, it should be higher up the pyramid of numbers.
(d) Dampness: so they do not dry out
Lack of light

Spinning a food web: Action

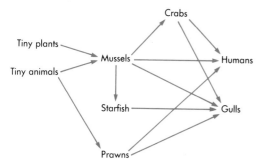

Predator v prey: Action

(a)

Predator	Prey
spider	fly
lioness	zebra
heron	fish
frog	worm

Growing plants from seeds: Action

Observations: five seeds had not germinated; some plants grow faster than others; there were two different plant varieties.

Inferences: the conditions were not the same over the whole tray.

Unit 3

Louis Pasteur: Did you know?

In A, soup in the neck prevents microbes in the air entering and turning the food bad. In B, microbes can get into the flask.

Amoeba: single-celled organism: Action

A moving; **B** feeding; **C** reproducing
Jennifer could set up a **choice chamber**. Some amoebas are put into a dish where half is illuminated and the other half is kept dark. After a while, the number of amoebas in each section is counted. The experiment is repeated several times to ensure that the results are not a fluke: to prove the hypothesis.

Food for thought: Action

(a) Brand X: it is much lower in fat and carbohydrate.
(b) Sugar;
(c) store in a chilled cabinet;
(d) Protein 5.0 g; Fat 3.0 g; Carbohydrate 23.0 g

Parts of a buttercup: Action

sepals 6; petals 6; stamens 9; carpels 1

Flower facts: Action

(a) Colour, smells and nectar attract bees and other insect which pollinate flower.
(b) The gardener brushes to transfer pollen from one flower to another. This ensures that the tomato flowers are pollinated.

Unit 4

All about blood: Action

(b) (i) A; (ii) England; (iii) Russian and Japanese people have similar distribution of blood groups.

The differences between twins: Action

The information suggests that height is largely inherited since there is little difference between twins growing up together or apart. Weight seems to depend more on the environment in which they were brought up.

Beetle features: Action

(a) Six legs; two antennae
(b) Length of beetle; length of antennae; pattern on back
(c) **Discontinuous variation** the beetles can be grouped according to their markings: no markings; spots; thin stripes; thick stripes.
Continuous variation length of the body; length of the antennae. You should try to measure these.
(d) Your way of recording the information may vary.

A peppering of moths: Action

(a) Dark-coloured moths would not be seen whereas light coloured would be. Predators would eat the light-coloured moths. Dark coloured moths would predominate.
(b) The black form had an advantage in city areas where they were camouflaged against blackened buildings. In country areas the black moths had no advantage.

A good likeness: Action

(a) Hoverfly
(b) Number of pairs of wings
(c) (i) A predator is unable to distinguish the honey-bee from the hoverfly. It knows from experience that the honey-bee stings. It is not prepared to take the risk of eating either creature.
(ii) Hoverflies with genes giving yellow/black striping had a natural selection advantage and so bred successfully.

Unit 5

Drawing bar charts: Action

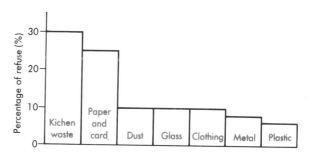

The effects of sulphur dioxide on seedlings: Action

(a) Seedlings grow better when sulphur dioxide is absent.
(b) Controls are needed to make sure the experiment is a fair test. The growth of the seedlings can be compared when only one variable is changed, i.e. the presence of sulphur dioxide.
(c) The experiment would no longer be a fair test. Light would be present in **A** and absent in **B**.

Pollution profile: Action

Z, X, W, Y

Fixing nitrogen: Action

(a) **Natural** lighting; bacterial action in soil and water
Unnatural oxides of nitrogen from vehicle exhausts and artificial fertilizers.
(b)

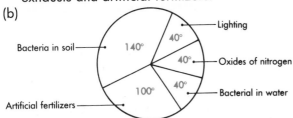

(c) **Naturally** 70 millions of tonnes
Unnaturally 110 millions of tonnes
(d) Perhaps with more vehicles and greater use of fertilizers the amount of non-natural nitrogen might increase. However, if we become more concerned with the environment we might find ways of reducing these levels.

The effects of detergents on growing plants: Action

(a) Temperature; light; volume of water in each beaker must all be kept constant.
(b) More plants die when the concentration of detergent increases.

Unit 6

Material concerns: Action

Object	Material	Property
A	concrete or cast iron	strength
B	Aluminium alloy or tubular steel	low density
C	Aluminium, copper or stainless steel	good conductor of heat
D	Aluminium alloy	low density
E	carbon fibre	strength

Stretching different types of 'wool': Action

(a) 54 cm; (b) **C** most, **A** least; (c) **C**

Testing the hardness of minerals: Action

A Hardness about 3; possibly calcite
B Hardness 7 or above; diamond, corundum, topaz, or quartz
C Hardness 4 to 5; apatite or fluorite
D Hardness 1; talc

Predicting the state of a substance: Action

Solids	Liquids	Gases
sodium	ethanol	hydrogen
iodine	mercury	nitrogen
sulphur	bromine	oxygen
zinc		ammonia
potassium chloride		
sodium chloride		
copper		
iron		

Solubility curves: Action

(a) (i) Sodium chloride; (ii) Potassium nitrate
(b) Sodium chloride
(c) 10 g
(d) 70 g of potassium nitrate crystallize out.
(e) 24°C

Solu-ability: Action

(a) Potassium nitrate
(b) Ammonia, hydrogen chloride
(c) Use litmus solution. The acid turns the litmus red and the alkali turns the litmus blue.

Testing for acids and alkalis: Action

A acid; **B** alkali; **C** neutral.

Measuring pH: Action

(a) It would be impossible to see the colour because of the strong colour of blackcurrant cordial.
(b) Bleach removes the colour from universal indicator.

Naming fractions: Action

Fraction 1 consists of ethanol and fraction 3 consists of water. Fraction 2 consists of a mixture of ethanol and water. If it is done carefully there will only be a small amount of fraction 2. It is difficult to separate completely a mixture of miscible liquids by fractional distillation.

Disproving forensic evidence: Did you know?

There could be other substances which behave in the same way as nitroglycerine. This is not quite enough evidence to ensure guilt.

Getting a reaction? Action

(a) Reaction:
copper (II) + zinc → copper + zinc sulphate

(b) No reaction
(c) No reaction
(d) Reaction:
copper (II) + magnesium → copper + magnesium
oxide   oxide
(e) No reaction

The acid test: Action

The hypothesis is only based on a few experimental results. She should test other metals, (e.g. iron, lead, copper etc). She might also try to keep variables constant, (e.g. temperature, concentration of acid etc).

Unit 7

Antoine Laurent Lavoisier: Did you know?

(a) The beam balance is very slow to use. It requires small weights to be added and taken off with tweezers. It is affected by air currents.
(b) (i) Mercury rises one fifth of the way up the bell jar.
(ii)

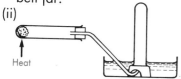

Heat

Joseph Priestley: Did you know?

(a) Oxygen; nitrogen; carbon dioxide.
(b) Fermentation; adding acid to chalk.
(c) Poor communications; language difficulties.

Rusting of iron uses up oxygen: Action

Set up apparatus as shown. Leave for a week. Water level rises one fifth of the way up the burette. Oxygen is used up which makes up one fifth of the air.

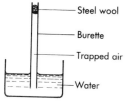

Steel wool
Burette
Trapped air
Water

Going a bit rusty: Action

(a) Air and water
(b) Air, water, salt and mud
(c) (i) Steel comes in contact with air and water.
(ii) Higher temperatures speed up the reaction.
(iii) Hot corrosive exhaust gases are found inside. These are in close contact with salt and mud on the road.

Trouble brewing: Action

Fermentation was not complete when the ginger beer was bottled. Fermentation produces carbon dioxide which builds up pressure inside the bottle until it explodes.

Resources in reserve: Action

(a) Tungsten
(b) Iron; chromium
(c) Aluminium
(d) They can be recycled or alternative materials (alloys) may be developed.
(e) Ceramics (based on clay) or glass.

Unit 8

John Dalton: Did you know?

(a) Colour blindness; weather; atomic theory
(b) He had insight into patterns etc but his experiments gave inaccurate results.
(c) It would have speeded up the progress of scientific understanding.
(d) It is important to test hypotheses carefully to check their truth. Do not discard them without evidence.

Structural problems: Action

iodine: molecules
sodium chloride: giant structures of ions
silicon dioxide: giant structure of atoms
water: molecules

Facing the elements, mixtures and compounds: Action

(a) 3; (b) 1; (c) 5; (d) 2; (e) 4

Changes of state: Action

(a)(i)   (ii)

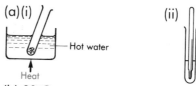

Hot water

Heat
(b) 90°C
(c) 78°C

Unit 9

Measuring rainfall: Action

(a)

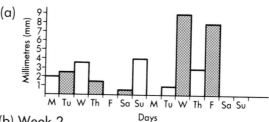

(b) Week 2
(c) Week 1: 2 mm
(d) Week 2: 3 mm
(e) Two weeks statistics are not enough to support this hypothesis. You should consider the results over many months.

Under pressure: Action

(a) Low: look at the pressures shown on the isobars
(b) Cold front
(c) 23°C
(d) Approx 11 mph and southerly direction

(e) Cloudy with sunny periods. As the cold front moves from the west, clouds will turn black (nimbus) and it will start to rain.

Exploring the –ites: Action

(a) Galena
(b) (i) Test with acid; (ii) Density
   (iii) Hardness or colour
(c) There is a difference in density. Sphalerite will float in a frothy detergent/water mixture whereas galena sinks.

Ages of rock: Action

1 melting;   5 heat and pressure;
2 cooling;   6 erosion;
3 erosion;   7 erosion
4 compacting;

Fossilized rocks: Action

Marble is formed under conditions which would destroy fossils

Water retention in soil: Action

(a) **Dry** soil should be broken down to the same size pieces.
(b) 10 seconds; 250 seconds; 60 seconds
   50 cm^3; 0.5 cm^3; 30 cm^3

Unit 10

Pulley forces: Action

(a) The forces are identical
(b) Repeat using different forces

Comparing strengths: Action

The bridge using folded paper.

Specific gravities: Action

(a) 1200 N; (b) 120 kg; (c) 200 N

Forced to weight: Action

(a) 50 N; (b) 500 N; (c) 10 000 N;
(d) 5 kg; (e) 50 kg; (f) 1 tonne

Chassis alloys: Action

The ideal material should be: cheap, readily available; easily moulded and shaped; low density (reduces fuel costs); able to hold its shape; strong (withstand impact); and not liable to corrode

Robert Hooke: Did you know?

(a) The amount of knowledge has become so great. There is a need to divide it up into different areas of study.
(b) **Advantage** Can study subject at great depth
   Disadvantage Unable to link information from different branches of science

Mean velocity: Action

40 mph

Travelling speeds: Action

(a) 42/7 = 6 m/s^2
(b) −24 m/s^2
(c) When the car stops suddenly, the passenger continues to move (because of inertia). Unless restrained, the passenger may hit the windscreen.

Putting into force: Action

(a) (i) The force directed forwards equals the backward directed force.
   (ii) The force directed backwards is greater. The passenger is pushed backwards into the seat.
   (iii) The force directed forwards is greater. The passenger is thrown forward.
(b) In a crash, passenger would be pushed backwards into their seat. Thus, injuries to head, body and legs are likely to be less severe.

Mass acceleration: Action

Acceleration = 300/100 = 3 m/s^2

Putting the brakes on: Action

(a) The time it takes for the driver to apply the brakes.
(b) It is directly proportional to the speed of the car.
(c) 12 m
(d) 9 m
(e) The velocity of the car is 20 m/s, so $v^2 = 400$. The braking distance is therefore 30 m.
(f) $KE = \frac{1}{2}mv^2$: the braking distance is directly proportional to KE.
(g) Friction between tyres and the road is reduced.
(h) $KE = \frac{1}{2}mv^2 = \frac{1}{2}m \times 400 = 200 \times m$
   This is converted into PE as the car travels up the ramp.
   $PE = mgh$
   $\qquad = 10\ mh$
   All of the KE is converted into PE.
   $200\ m = 10\ mh$
   $h = 20$ metres

Unit 11

Pupil thinks quickly: Action

The battery could short circuit.
The heat produced could ignite fuel.

Completing a circuit: Action

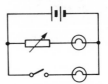

Realizing potential: Action

(a) V, voltmeter; A, ammeter
(b) To adjust the current flowing so that different sets of results can be taken.

(c)

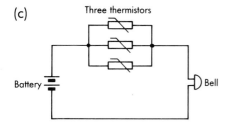

(d) Reading 1.5 A, 2.8 V: it deviates from the straight line.

(e) As voltage increases, current increases in direct proportion.

Ring-mains: the wires and wherefores: Action

(b) Either switch will turn the bulb on or off.

(c) At the top and bottom of the stairs. It allows switches to work without the irritating (or perhaps difficult) task of walking up or down the stairs.

Motor cars: Action

Battery: car; milk float
Overhead cables: tram
Electric track: tube train

Watts the charge? Action

£866

The science of appliance: Action

(a) Electric clock
Light
TV
Hair dryer
Iron
Heater

(b) Electric clock, one thirteenth of a unit
Light, one tenth of a unit
TV, one eighth of a unit
Hair dryer, one half of a unit
Iron, one unit
Heater, two units

(c) (i) 5p; (ii) 25p; (c) 10p

Handle with care: Action

A The flex acts as a trip-wire.
B The electric socket over the sink is dangerous. Water and electricity combines can cause fatal shocks!
C Poking a fork into a toaster could lead to an electric shock.
D An electric cable is placed over a cooker. The cable could burn.
E Multiadaptors can lead to overloaded circuits.

Unit 12

Setting alarm bells ringing: Action

(a) Bell stops ringing if wires burn through. This **is** a disadvantage.

(b)

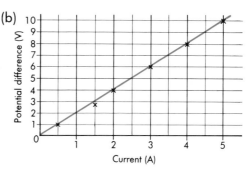

Tabling the truth: Action

(a)

Input		Output
S_1	S_2	
0	0	0
0	1	1
1	0	1
1	1	1

(b) OR

(c)

Seeing the light: Action
5, 5, 4, 5, 6, 3, 7, 6

Unit 13

James Prescott Joule: Did you know?

(a) Water at the bottom of the waterfall has gained KE as it falls from top to bottom.

(b) Hotter

Energy changes: Action

(a) Mechanical (or kinetic) energy into electrical energy.
(b) Mechanical (or kinetic) energy into heat (and sometimes sound).
(c) Chemical into heat, mechanical energy and sound.
(d) Electrical energy into heat and light.
(e) Chemical energy into heat and light.

Power crazy: Action

(a) (i) 10%; (ii) 59%; (iii) 3%
(b) 28%
(c) Low efficiency.

The heat of the moment: Action

A Conduction; **B** Convection; **C** Convection;
D Radiation; **E** Conduction

Fuel for thought: Action

(a) Methylated spirit produces more energy per gram. (Larger temperature rise even though only half the quantity was burnt.)
(b) Fuel B is more economical.

Fuelling the energy debate: Action

(a)
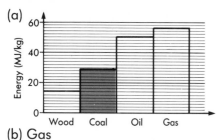

(b) Gas

Hydro-generation: Action

(a) The West of England, Wales and the West of Scotland
(b) These are hilly areas
(c) There is a higher rainfall than in the east.
(d) No. Holland is a very flat country.

The Swedish solution: Action

(a) Conifer forests have a wide range of other uses such as paper making, furniture etc.
(b) They grow quickly; they have a low nitrogen and sulphur content; and, on burning, produce an ash rich in potassium.
(c) It produces less air pollution on burning
(d) Fertilizer
(e) Photosynthesis

GB sources: Action

(a) (i) Oil; (ii) Hydroelectric; (iii) Hydroelectric; (iv) Coal
(b) Less oil would be used due to its cost and so more coal might be burned to provide the energy required.

Unit 14

How quickly does sound travel through the air? Action

speed of sound = distance/time
$$= 1000/3$$
$$= 333 \text{ m/s}$$

Taking soundings: Action

(a) The sound travels to the cliff and back. If the distance between the ship and cliff is x m, sound travels $2x$ m.
$$2x = 300 \times 5 = 1500 \text{ m}$$
$$x = 750 \text{ m}$$

(b) Distance travelled by sound = 165 m
Time taken = 0.5 s
Velocity of sound = 165/0.5 = 330 m/s

The assumption is that the sound takes no time travelling through the amplification system.

Speed of sound through different materials: Action

(b) The speed of sound increases as the density increases.

(c) Very slow, because polystyrene has a very low density.
(d) 1200 m/s: Softwood is slightly less dense than water.
(e) Red indians want to know when the train is coming so they can ambush it. Because sound travels almost twenty times faster through steel than through the air, they hear it first through the rail.

Getting to the pitch: Action

The slide is pushed in and out so altering the length of the column of vibrating air. When the slide is out, the tube is long and a deep note is produced.

Unit 15

Shadows: Action

A large shadow is produced when the object (mug) is close to the lamp and when there is a large distance between object and screen.

Reflections in a mirror: Action

A B C D E H I K M O Q T U V W X Y
c i l m o v w x

Portraying an image: Action

(a) 10 cm
(b) 2 cm
(c) The picture would be out of focus.

In camera: Action

Fig.1 shows the camera set up using the viewfinder. The light ray entering the camera passes through a series of lenses. The light ray is reflected by a plane mirror and through a five-sided prism.

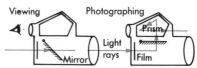

When the picture is ready to be taken, the plane mirror is folded back. The light ray now strikes the film.

The advantage of this type of camera is the fact that the view through the viewfinder is exactly what is taken. This is not the cases with other cameras where the photographer gets a slightly different view through the viewfinder.

The eyes have it: Action

(a) Two eyes at the sides of the head give better all round view.

(b) Two eyes at the front of the head (e.g. human) has the advantage of better 3−0 vision since each eye has a slightly different view of an object.

Lens in a box: Action

(a) B; (b) D; (c) A; (d) A

Unit 16

Ordering orbits: Action

(a) (i) Mercury;
(ii) Mercury and Venus;
(iii) Earth;
(iv) Jupiter

(b) Mercury; Venus; Earth; Mars; Jupiter; Saturn; Uranus; Neptune; Pluto

(c) Planets close to the Sun have smaller diameters and larger densities than those planets furthest from the Sun.

Planet X: Action

A Jupiter
B Venus
C Earth

Plutonic poser: Action

It is believed that there might be a tenth planet. The gravitational force exerted by this planet alters the path of Pluto. So sure are scientists that this planet exists they have named it Humphrey.

The Earth as the centre of the Universe: Did you know?

(a) Ptolemy's theory left room outside fixed stars for heaven and hell. It did not conflict with the Bible.

(b)(i) The moon has much larger mass than the spacecraft. Therefore, the pull on spacecraft is much larger than the pull on the Moon.

(ii) Planets of heavy mass exert a strong force of mutual attraction.

The planet Saturn: Did you know?

(a) Huygens had a better telescope.
(b)

1610 and 1613   1612

The Earth is a sphere; not a disc: Did you know?

If the Earth was a disc the shadow would be elongated and not circular.

The sky's the limit? Action

Since light takes such a long time to reach the Earth, scientists can only observe the light that left the star ages ago!

History of the Universe: Did you know?

(a) The Hubble telescope was launched in 1990. It is a powerful telescope outside the Earth's atmosphere. It will give better pictures of outer space.

(b) There is still room left for God as creator at the time of 'big bang'. Also, the church wanted to avoid the sort of problems which existed with Galileo.

INDEX

Please note that this index is complementary to the contents list and glossary—with which it should be used.

174

USEFUL ADDRESSES

Here are some addresses to help you to find out information for researching the Project activities.

There are **many** other useful contacts that you will need to discover for yourself.

British Nuclear Fuels Ltd
Information Services BNFL
Sellafield
Cumbria
CA20 1PG

Campaign for Nuclear Disarmament (CND)
22-24 Underwood Street
London
N1 7JQ

Christian Aid
Interchurch House
35 Lower Marsh
London
SE1 7RT

Christian Ecology Link
17 Burns Gardens
Lincoln
LN2 4LJ

Friends of the Earth
377 City Road
London
EC1V 1NA

Global Education Network
6 Endsleigh Street
London
WC1H 0DX

Greenpeace UK
30-31 Islington Green
London
N1 8XE

Intermediate Technology Development Group
Myson House
Railway Terrace
Rugby
Warwickshire
CV21 3HT

Junior Astronomical Society
c/o The Secretary
36 Fairway
Keyworth
Nottingham
NG12 2DU

National Centre for Alternative Technology
Llwyngwern Quarry
Machynlleth
Powys
SY20 9AZ

Nature Conservancy Council (NCC)
Northminster House
Northminster Road
Peterborough
Cambridgeshire
PE1 1UA

OXFAM
274 Banbury Road
Oxford
OX2 7DZ

Tearfund
100 Church Road
Teddington
Middlesex
TW11 8QE

World Wide Fund for Nature (WWF)
Panda House
Weyside Park
Godalming
Surrey
GU7 1XR

Y Care International
640 Forest Road
London
E17 3DZ

Your local Member of Parliament
House of Commons
London
SW1A 0AA